Experimental Crystal Physics

W. A. WOOSTER *and* A. BRETON

Experimental Crystal Physics

SECOND EDITION

CLARENDON PRESS · OXFORD
1970

Oxford University Press, Ely House, London W.1

GLASGOW NEW YORK TORONTO MELBOURNE WELLINGTON
CAPE TOWN SALISBURY IBADAN NAIROBI DAR ES SALAAM LUSAKA ADDIS ABABA
BOMBAY CALCUTTA MADRAS KARACHI LAHORE DACCA
KUALA LUMPUR SINGAPORE HONG KONG TOKYO

FIRST EDITION 1957
SECOND EDITION 1970

PRINTED IN GREAT BRITAIN
BY SPOTTISWOODE, BALLANTYNE & CO. LTD.
LONDON AND COLCHESTER

Preface to the Second Edition

THE physical properties of solid crystalline materials have become more important with the passage of time. The applications of the properties of single crystals of quartz, silicon, germanium, and many more complex solid structures have become widespread both in scientific, technical, and even domestic fields. The study of the physical properties of these materials is not easy and the experiments described in this book have been designed to bring the student into contact with the materials he is studying. The symmetry of crystals and the properties of tensors are two of the important aspects of this work which these experiments are designed to illustrate. The experiments have been deliberately made simple so that they can be performed within a period of two hours. As far as possible the apparatus has been made to show the principles of the experiment rather than to produce a highly accurate result.

The 'classical' experiments were originally developed in the Department of Mineralogy and Petrology of Cambridge University, England. The present design is due to Mr. Geoffrey A. Wooster and the construction is by Crystal Structures Ltd., Bottisham, Cambridge, England. The experiments that have been added since the first edition of this book were developed in the Department of Solid State Physics, École Polytechnique, Montreal under the guidance of one of the authors, while he was teaching there. Some theory has been given to introduce each experiment but this has been limited so that a student could read it without taking up too much laboratory time.

The references to published work are not intended to be complete but may prove useful for further study.

Cambridge W.A.W.
Montreal 1969 A.B.

Contents

1. Optical Properties of Crystals

1.1. Introduction

THE first work on the double refraction of crystals was done in the seventeenth century by Bartolinus of Bruges. Some decades later Huyghens accurately described the phenomena to be seen in calcite and Hooke gave an almost modern account explaining the effects as being due to molecules shaped like ellipsoids of revolution. During the nineteenth century Fresnel and others developed theories of the propagation of light through transparent media and found some difficulty in relating the known properties of doubly refracting crystals to their theories. Maxwell developed his electromagnetic theory of light towards the end of the nineteenth century and although this is still regarded as a completely satisfactory theory when applied to crystals, he did not present the simplest account of the optical properties of crystals. A mineralogist, Lazarus Fletcher, who was keeper of minerals in the British Museum of Natural History, London, put forward his account of the optical indicatrix in 1891 and even today there is no more comprehensive nor any simpler treatment of this subject.

1.2. Definition of the indicatrix

In general, for crystals of low symmetry, the indicatrix is a triaxial ellipsoid having its principal axes of length proportional to α, β, and γ, the three principal refractive indices of the crystal. Crystals belonging to the tetragonal, rhombohedral, and hexagonal systems are represented by indicatrixes that are ellipsoids of revolution. The axis of revolution coincides with the tetrad, triad, or hexad axis of the crystal. The indicatrix of a cubic crystal is a sphere of radius proportional to the refractive index.

A radius vector of the triaxial indicatrix has a length proportional to the refractive index of the light that *vibrates* in the direction of that radius vector. According to the electromagnetic theory of light the vector representing the alternating electric field lies in the plane of the wave front. This vector is known as the vibration direction and is normal to the direction of propagation of the light wave. Thus the wave normal is perpendicular to the vibration direction. The following construction enables the wave normal associated with a particular vibration direction to be found. In Fig. 1.1, OA represents a vibration direction drawn from the centre of the ellipsoid O to the surface at A. The line AN is normal to the surface of the ellipsoid at A and ONR is a line perpendicular to AN. OR represents the direction in which the ray of light, having OA as vibration direction, is transmitted. The line OW, perpendicular to OA and lying in the plane defined by OA and AN is the wave-normal direction. The refractive index of the light having OW as wave-normal direction and OA as vibration direction is proportional to OA. The velocity of the ray OR is proportional to I/AN.

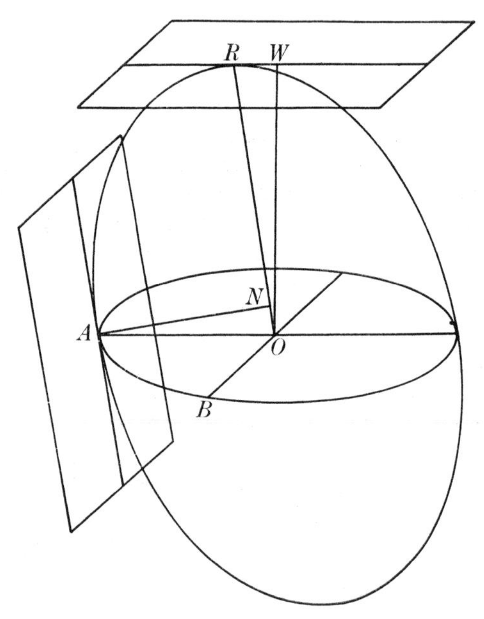

FIG. 1.1. The triaxial ellipsoid representing the optical indicatrix. OW represents a wave normal perpendicular to the tangent plane at R and the parallel central section AOB. AO, BO are the principal axes of this central elliptical section. AN is perpendicular to the tangent plane at A and to the ray OR, corresponding to the wave normal OW.

1.3. The physical basis for the indicatrix construction

When any kind of transverse vibration—it may be purely mechanical or optical or wireless waves—enters an anisotropic medium, it is resolved into two components having mutually perpendicular vibration directions. These vibration directions correspond to the maximum and minimum restoring forces acting upon the displaced electric charges produced by the wave. For a crystal of low symmetry the variation of the refractive index with direction of vibration is found experimentally to be represented by a triaxial ellipsoid and not by some more complicated figure. When light travels along a given direction OW (Fig. 1.1), the vibration directions must, from what has been said above, be parallel to the principal axes of the elliptical section of the indicatrix OA, OB, to which the wave normal is perpendicular. In the directions OA and OB there is no change of refractive index with a small change in the vibration direction. It is characteristic of all such transverse vibrations that this determines the vibration directions.

The wave front normal to OW is a tangent to the indicatrix at the point R. The tangent plane to the surface of the indicatrix at A is parallel to OR and the normal to the indicatrix at A intersects OR at N. This result is a property of a triaxial ellipsoid and can be proved by the usual geometry. It is also a property of this ellipse AR that

$$AN.OR = \text{constant} = k.$$

If now the velocity of the ray, v, corresponding to the wave normal OW, is made proportional to $1/AN$, i.e.

$$v = k/AN$$

then

$$OR = v.$$

Thus the line OR is of a length proportional to the velocity of the ray in that direction. The geometry of the figure requires that OW, OR, OA, and AN shall all lie in the same plane. A second ray direction corresponding to a wave normal OW is obtained by drawing the normal to the indicatrix at point B. The ray lies in the plane BOW and is perpendicular to the normal through B.

1.4. Optic axes and the indicatrix

Every triaxial ellipsoid has two circular sections, Fig. 1.2. These are shown passing through the ends of the β-axis and points A, B, intermediate between α and γ where the radius vector is of length β. Normal to these circular sections are the optic axes marked $O.Ax$. Along the optic axes light waves travel with no double refraction. All radii of the circular sections are equal to β and there is therefore no double refraction of the light. The acute angle between the optic axes in the crystal is called $2V$. Round each optic axis there are directions for each of which the difference of refractive index for the two beams of light associated with a particular direction is constant. The surface on which such wave normals lie is very nearly a cone when it is of small angle. Two such cones are shown in Fig. 1.3. As the double refraction increases the figures become more complicated than cones, as may be seen in convergent light pictures.

1.5. Observations between crossed Nicols or polaroid filters

When polarized light falls upon a plate of a doubly refracting crystal, the vibration direction of the light is resolved into components parallel to the major and minor axes of the central section of the indicatrix to which the plate is parallel. In Fig. 1.4, OP represents the vibration direction of light travelling upwards through the paper and AO and OB the components into which this vibration is resolved in the directions which are possible

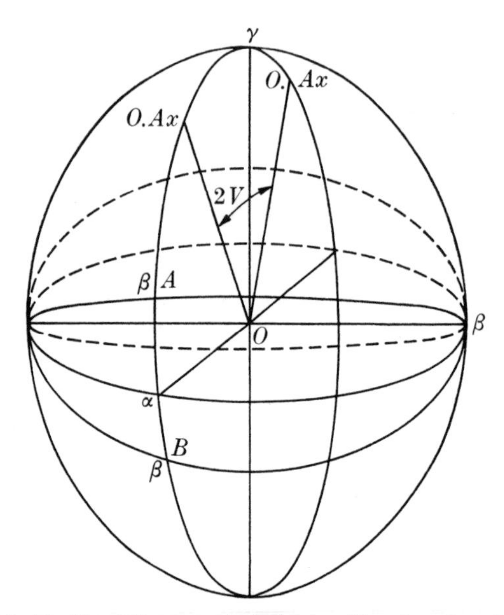

FIG. 1.2. A triaxial ellipsoid having principal axes of length α, β, and γ respectively. $AO\beta$ and $BO\beta$ are circular sections of the indicatrix and the optic axes, denoted ***O.Ax***, are normal to them.

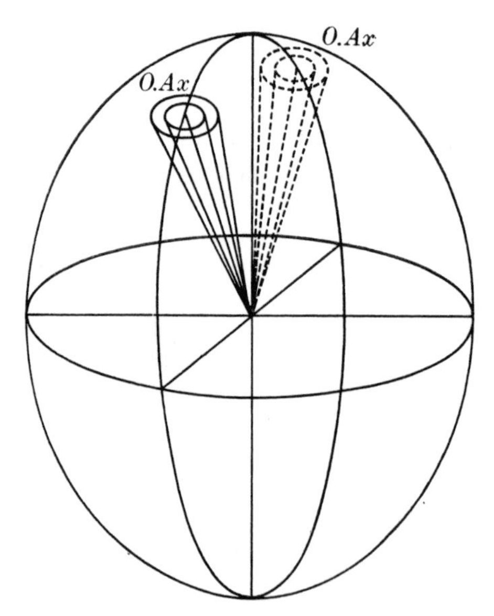

FIG. 1.3. Cones drawn round the optic axes to represent directions for which the difference in path for the doubly refracted waves is λ, 2λ, etc. for a given distance travelled in the crystal.

vibration directions in the crystal. If OP is of unit amplitude and $\angle AOP = \theta$ then the components are

$$OA = \cos\theta, \qquad OB = \sin\theta.$$

In travelling through the crystal the refractive indices associated with OA or OB are n_1 and n_2 respectively. If the thickness of the crystal plate is t the equivalent path in air for the vibration along OA is $n_1 t$ and for the

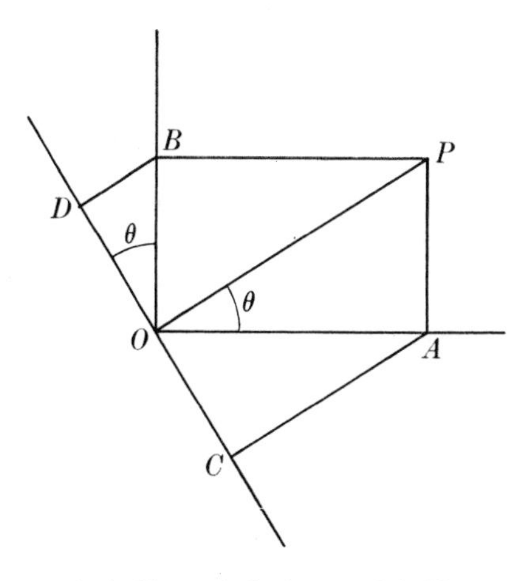

FIG. 1.4. OP represents the incident polarized wave; OA, OB represent the components into which the crystal resolves it; OC, OD represent the amplitudes of the components transmitted by the analyser.

other vibration it is $n_2 t$. There is thus a path difference of $t(n_1 - n_2)$ and the phase of the vibration OA differs from that of OB by angle ϕ where

$$\phi = \frac{2\pi}{\lambda} t(n_1 - n_2).$$

The analyser is crossed with respect to the polarizer and the amplitudes of the two vibrations are given by the equal components OC and OD in the line COD which is perpendicular to OP. If the phase of the vibration OD be taken as zero, that of OC is $(\pi + \phi)$. The emerging light therefore has the amplitude ξ where

$$\xi^2 = OD^2 + OC^2 - 2.OD\,OC\cos\phi.$$

From Fig. 1.4 it may be seen that

$$OC = OD = \sin\theta\cos\theta = \tfrac{1}{2}\sin 2\theta.$$

Hence,

$$\xi^2 = \sin^2\theta\cos^2\theta(2-2\cos\phi)$$

$$= \tfrac{1}{2}\sin^2 2\theta . 2\sin^2\frac{\phi}{2},$$

$$\xi = \sin 2\theta \sin\frac{\phi}{2}.$$

Thus as a crystal plate is rotated relative to the crossed polaroids $\sin 2\theta$ become zero when θ is 0°, 90°, 180°, 270°, and is a maximum when θ is 45°, 135°, 225°, 315°. Further, ξ is zero when ϕ is 0 or 2π or multiples of 2π. Thus if a plate has a given thickness, ϕ increases steadily with the inclination of the light to the plane of the plate. A uniaxial plate cut perpendicular to the optic axis has directions lying in cones about the normal to the plate for which ϕ is 2π, 4π, 6π, etc. and for each of these $\xi = 0$. Thus between crossed polaroids such a plate gives a series of concentric dark and light circles when viewed in convergent light (see 1.6.4).

1.6. Experimental methods

A complete description of the optical properties of transparent colourless crystals is given by the three principal refractive indices and their orientation with respect to the external shape of the crystal. The experiments described in the following paragraphs show how the refractive indices, and their differences can be found. The differences are important because many of the phenomena shown by anisotropic crystals depend more on the differences in refractive indices—the double refractions—than upon the absolute value of the indices. Optically active crystals form a class by themselves and have to be studied by special methods, some of which are described in this chapter. The properties of opaque crystals may be studied using reflected light. The full treatment of these properties is beyond the scope of this book but an introduction is given by the experiments on cubic crystals.

1.6.1. *Measurement of double refraction using a prism refractometer*

1.6.1.1. *Principle of the method*

When light passes through a piece of transparent solid bounded by two faces making an included angle A with one another, the light is deviated through an angle D. The value of D depends on the angle at which the light falls on the first face of the prism formed by the inclined faces. When within the prism the light travels in a direction equally inclined to the two faces, the deviation is a minimum and the relation of A and D to the refractive index n, is given by the equation

$$n = \sin\frac{A+D}{2}\Big/\sin\frac{A}{2}.$$

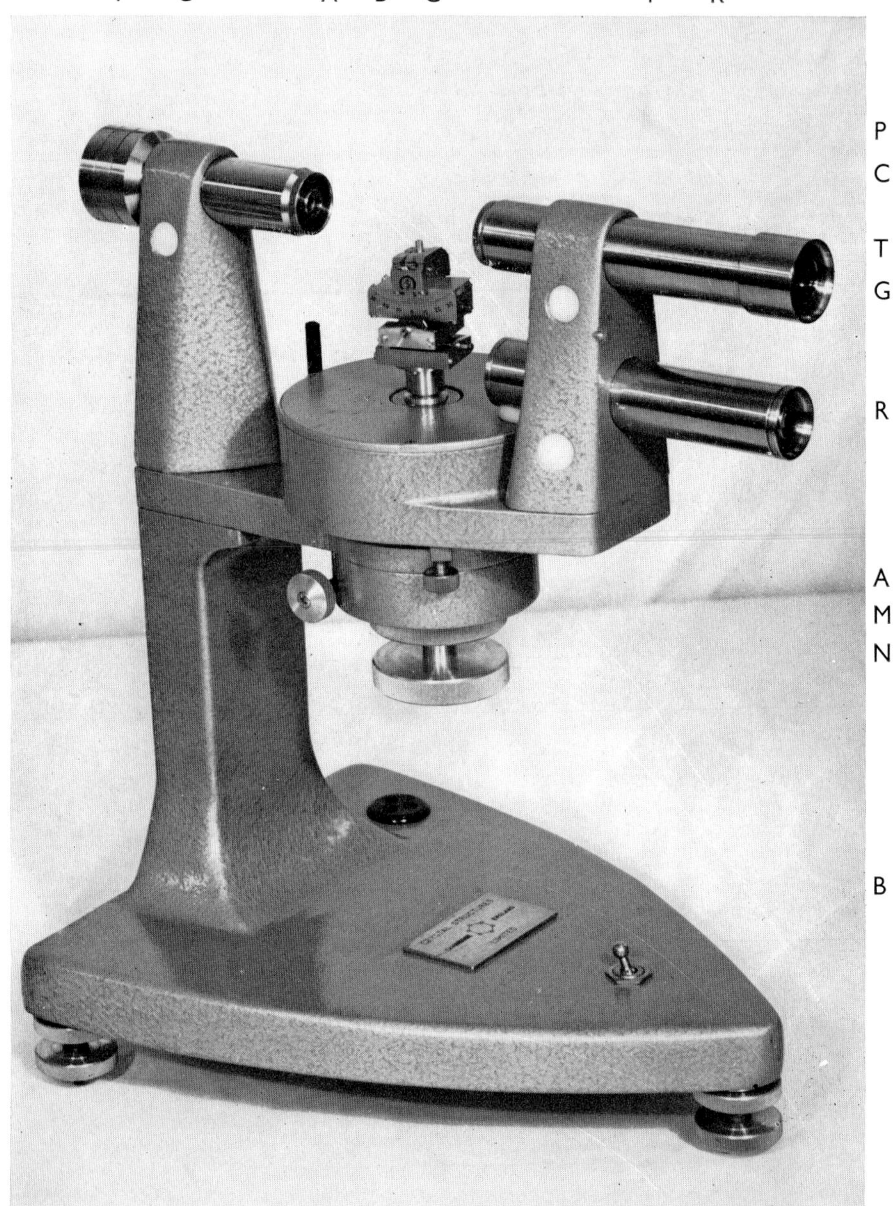

FIG. 1.5. Single-circle optical goniometer suitable for use in measuring the refractive indices of a crystalline prism. (A given letter is used twice to give coordinates of a part of the instrument denoted by that letter.)

Since the maximum value of $\sin\frac{1}{2}(A + D)$ is unity there is a limiting value of A equal to $2\sin^{-1} 1/n$.

In a doubly refracting crystal there are in general two beams of light transmitted by such a prism, one more deviated than the other. In a uniaxial crystal, such as quartz or calcite, the two beams correspond to the principal refractive indices when the light travels in the crystal in a direction perpendicular to the optic axis.

For these two crystals this axis is the crystallographic trigonal axis. Thus the condition for being able to observe the two principal refractive indices, which are known as the ordinary and extraordinary refractive indices respectively, is that the plane bisecting the angle A between the two faces contains the directions of the optic axis. The inclination of the edge where the two faces meet to the optic axis is irrelevant. Whatever this angle may be the light travels in the crystal perpendicular to the optic axis. The two beams travelling through the crystal are polarized; the vibration direction of the ordinary ray is perpendicular to the optic axis of the crystal and the vibration direction of the extraordinary ray is parallel to that axis. On emerging from the crystal the two beams travel in different directions and their directions of polarization are mutually perpendicular.

1.6.1.2. *The optical goniometer*

For this experiment the instrument, shown in Fig. 1.5, is provided with a standard goniometer head, G, for holding the crystal, mounted on the axis fixed to the single horizontal glass scale divided into $\frac{1}{3}°$ intervals. An image of the scale is seen in the reading microscope R which is provided with a vernier giving $1'$ readings. The lamp illuminating the scale is provided with current from a transformer under the base B. There are two knurled nuts N and M on the lower end of the shaft carrying the goniometer head and scale. N is permanently fixed to this shaft and serves to rotate the crystal freely through a large angle. A fine adjustment can be made by clamping the nut M to the shaft and turning the tangent screw A. A Websky slit covers the outer end of the collimator C. If light from a mercury lamp is being used a colour filter, P, orange, green, or violet, is also inserted here. At the inner end of the collimator a standard 3-in or 75-mm lens is screwed in. The telescope T has a Ramsden eye-piece and a graticule. There is no fine adjustment for the setting of the telescope. Two polaroid filters may be used as polarizer and analyser respectively. The polarizer fits on to the collimator tube between the lamp and the Websky slit. The analyser may be placed over either end of the telescope tube.

1.6.1.3. *Experimental procedure*

The telescope is directed towards a distant object and the eyepiece is moved until this object is clearly seen. The graticule should also be clearly seen and there should be no relative movement of the image of the distant object and the graticule, when the observer moves his eye from side to side of the eyepiece, i.e. no parallax. The telescope is then set in line with the collimator and the Websky slit is illuminated. The image of the slit should be seen sharply focused on the graticule and there should be no parallax between them. If this condition is not fulfilled the Websky aperture must be moved until it is.

The crystal is mounted on the goniometer head. If it is large enough it may be pushed into a small piece of plasticine on the top of the arcs. If it is too small for this it may be stuck to a glass or metal piece which can then be mounted in plasticine on the goniometer head. It is important to mount the crystal with one face of the prism to be used as nearly as possible perpendicular to the axis about which the uppermost arc turns. If this is done an adjustment of that arc does not change the direction of the normal to that face of the prism.

The lower arc is adjusted until the reflection from the face just referred to gives an image of the Websky aperture which is central with respect to the intersection of the cross lines. The goniometer head is rotated on its vertical shaft by turning the wheel N and the reflection from the second face is sought. It is usually easier to see this by the naked eye than through the telescope until the crystal is almost correctly adjusted. Using only the upper arc this reflection is brought into a horizontal plane. It is then possible to pick up the reflection in the telescope. When the crystal is correctly adjusted the images formed by reflection from both faces should appear central in the telescope as the crystal is rotated about the main vertical axis. (The telescope is kept clamped at any convenient angle to the incident beam during the making of this adjustment.)

The refracted beams are now sought. It may be easier to pick them up in the first instance by the naked eye. If white light is used they are distinguished from the reflections from the prism faces by their spectral colours. If monochromatic light is in use they can be distinguished from surface reflections by the fact that they can be extinguished in turn by rotating the polaroid filter. For the final measurement the mercury lamp is used with one of the three filters giving orange, green, or violet light. When a refracted beam is seen in the telescope the crystal is rotated about a vertical axis, by turning the wheel N, and at the same time the telescope is moved. The direction of movement is towards a smaller angle of deviation. Finally it will be found that whichever way the crystal is rotated the image of the slit moves towards positions corresponding to greater angles of deviation. The crystal is clamped in the setting giving minimum deviation, by turning the knurled wheel M. The telescope is now moved so that the intersection of the cross lines in the graticule coincides with the centre of the image of the Websky aperture. The angle on the scale is read. Let this be θ_1. Keeping the crystal fixed the telescope is now turned to pick up the direct beam and, when the image of the slit is at the centre, the angle of the scale is again read. Let this be θ_2. Then the angle of minimum deviation D is given by

$$D = \theta_1 - \theta_2.$$

The same sequence of settings is used to find the angle of deviation corresponding to the other refracted beam. The setting of the crystal is necessarily different in the positions of minimum deviation for the two beams.

The angle A between the faces forming the prism is found in the following way. The telescope is clamped in any position inclined to the incident beam, for example 90°, and the crystal is rotated about the vertical axis until a reflection from one of the two faces is observed. Let the scale reading be θ_3. The crystal is now turned until the other prism face gives a reflection at the centre of the eyepiece field. Let the scale reading in this setting be θ_4. Then

$$A = \pi - (\theta_3 - \theta_4)$$

since A is the included angle and $(\theta_3 - \theta_4)$ is the angle between the face normals.

The vibration directions of the two refracted beams may be determined by using the analyser which, in Fig. 1.5, is shown slipped over the eyepiece of the telescope. As the analysing polaroid disc is rotated on the telescope barrel the refracted image is extinguished twice during a complete turn. At the setting for extinction the vibration direction of the refracted beam is perpendicular to that of the light transmitted by the analyser. The vibration direction of the analyser may be found by looking at the reflection of daylight from a polished glass or wooden surface. When the glancing angle of reflection is about 35° it will be found that on rotating the analyser the intensity of the light transmitted through it rises and falls. The vibration direction of the light reflected from the polished glass is mainly parallel to the glass surface. When the polaroid most reduces this reflection its vibration direction lies in a plane perpendicular to the plane of the glass.

The vibration directions of the refracted beams should be related to the direction of the principal axis (trigonal axis for calcite) of the crystal.

EXAMPLE

A crystal of calcite showing the faces of the scalenohedron $\{21\bar{3}1\}$ was used. It was set up on the goniometer so that the intersection of the two faces $(3\bar{1}\bar{2}1)$, $(\bar{3}211)$ was vertical and parallel to the axis of the goniometer. The angle between the normals to the faces was found to be 132° 58′. Hence the angle A of the prism is 47° 2′.

Using sodium light the position of minimum deviation for one refracted beam was determined. The vibration direction of this light was found using a sheet of polaroid on which had been scribed the direction of vibration of the light it transmitted. It was parallel to the optic axis of the crystal and the light was, therefore, the extraordinary beam. The minimum deviation was given by the following readings:

$$\text{direct beam} = 56°\ 10',$$
$$\text{deviated beam} = 81°\ 52',$$
$$\text{minimum deviation } D = 25°\ 42'.$$

Hence

$$n_e = \frac{\sin \frac{1}{2}(A + D)}{\sin \frac{1}{2}A} = 1{\cdot}486.$$

The second refracted image was found to have a vibration direction perpendicular to the trigonal axis and corresponded, therefore, to the ordinary beam. The corresponding readings for the minimum deviation were

$$\text{direct beam} = 56^\circ\ 10',$$
$$\text{deviated beam} = 92^\circ\ 0',$$
$$\text{minimum deviation } D = 35^\circ\ 50'.$$

Hence

$$n_0 = 1{\cdot}659.$$

(Kaye and Laby, *Physical and Chemical Constants*:

$$n_0 = 1{\cdot}658,\ n_e = 1{\cdot}486.)$$

1.6.2. *The measurement of double refraction by means of total reflection*

1.6.2.1. *Principle of the method*

When a beam of light travelling in a medium of refractive index n_1 strikes a plane boundary at an angle of incidence i it is, in general, partially reflected from, and partially transmitted into, the medium of refractive index n_2, on the other side of the bounding plane. If the angle of refraction is r then by Snell's Law

$$\frac{\sin i}{\sin r} = \frac{n_2}{n_1}.$$

Thus if n_1 is greater than n_2, r is greater than i. But $\sin r$ cannot exceed unity it follows that r cannot be greater than 90°. The value of the angle i for which $\sin r$ is unity, namely $\sin^{-1} n_2/n_1$, is called the critical angle and denoted by I_t. If i exceeds the critical angle the light is not refracted at all but only reflected, i.e. it is totally reflected.

To utilise this total reflection in measuring refractive indices a polished crystal plate is immersed in a liquid of refractive index higher than its own index. The light reflected at different angles from the crystal when immersed in the liquid is observed in a telescope. When the crystal is set relative to the telescope so that the angle between the normal to the plate and the axis of the telescope is I_t half of the field of view of the eyepiece is brightly illuminated because of the total reflection and the other half is less bright. There is a sharp boundary separating the two fields. This observation can be made with the light reflected to the right or to the left of the incident beam. In both settings the angle between the normal to the reflecting surface and the axis of the telescope is I_t. In going from one setting to the other the normal to the plate must be turned through an angle $2I_t$ relative to

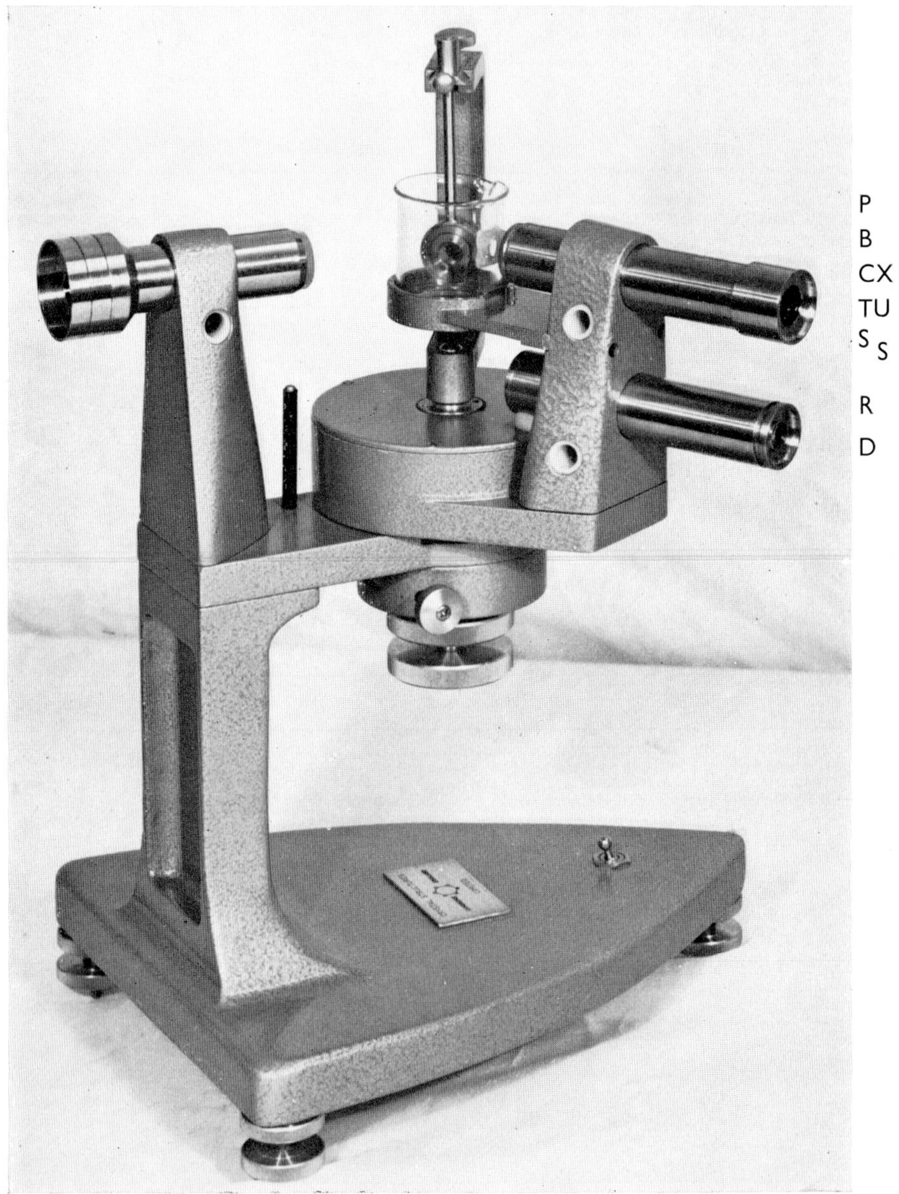

Fig. 1.6. Kohlrausch refractometer suitable for use in measuring the refractive indices of a crystal plate by means of total reflection.

the telescope. Thus the observation of this angle $2I_t$ and the refractive index of the liquid enables the refractive index of the crystal to be found.

A doubly refracting crystal plate gives, in general, two boundaries each separating a bright field from an adjoining less bright field. Thus, for example, a plate cut parallel to the optic axis of a uniaxial crystal gives a separation of the two boundaries that depends on the orientation of the optic axis relative to the plane of reflection. When the optic axis lies in the plane of reflection the light refracted at an angle $r = 90°$ is travelling along the optic axis and both beams have the same refractive index. In this case there is only one total reflection boundary. If the plate is turned in its own plane through 90° the optic axis is perpendicular to the plane of reflection. In this case light for which $r = 90°$ is travelling in a direction perpendicular to the optic axis. There are two refractive indices, n_o and n_e, corresponding to the ordinary and extraordinary waves respectively. Two boundaries correspond to these refractive indices, one being in the same position as was given in the first orientation of the plate.

Since all the light is reflected in the brightest part of the field of view it is not polarized. In the region between the two boundaries one beam is partially transmitted and the other totally reflected. The light in this region is therefore polarized. The direction of polarization depends on whether the ordinary or the extraordinary beam has the higher refractive index.

1.6.2.2. *The total reflection refractometer*

Figure 1.6 shows an apparatus adapted for the measurement of the angle of total reflection. The collimator tube C has no slit because light must fall on the crystal over a range of angles greater than that corresponding to the field of view of the telescope. A 32-mm focal length lens can be with advantage screwed into the end of the collimator tube near the crystal. Monochromatic light is essential. This may be achieved by using a mercury lamp and a filter mounted in the end of the collimator tube. The crystal plate, X, is mounted on a disc that can be rotated about a horizontal axis when it is desired to change the orientation of the optic axis (which is parallel to the plane of the plate) relative to the (horizontal) plane of reflection. The support for this disc is carried on a rod P, which may be clamped at a suitable height in a G-shaped metal piece. This G-shaped support is carried on a vertical shaft which is rigidly attached to the graduated glass scale which is inside the box D. This scale is divided into $\frac{1}{3}°$ intervals and may be read by the microscope R which is provided with a vernier reading to 1′. A metal tray, U, is fixed to the vertical support of the telescope, T, by the screws S, S. Fitting into the tray is a glass vessel, B, which has a parallel-sided optically flat window permitting a parallel beam reflected from the crystal to remain parallel on entering the objective of the telescope. This window must be arranged to be just opposite the objective of the telescope. The vessel B is filled with a high refractive index liquid up to a level just above the top of the window. A polarizing filter may be used either between the lamp and the crystal or between the crystal and the observer. In Fig. 1.6 the polarizing filters are shown in both places but only one is required in this observation.

1.6.2.3. *Experimental procedure*

The telescope should be adjusted so that a distant object is clearly focused on the cross lines and there should be no parallax between the image of that object and the cross lines. This adjustment has been made by the makers but can be checked by the observer moving his eye from side to side across the eyepiece and noting whether there is any relative movement of the images. The lamp and filter are arranged so as to send a monochromatic beam down the collimator. The temperature of the liquid is read at the beginning of the observation and also at the end because the refractive index of the liquid changes rapidly with the temperature. Table 1.1 shows the refractive indices of α-monobrom naphthalene at various temperatures. The telescope is moved with the crystal and the reflected light observed in the telescope. The sharp boundary or boundaries can be brought to coincide with the intersection of the cross lines in the eyepiece. The angle is read in R. To avoid confusion when there are two boundaries it is convenient to extinguish one of them using the analysing polaroid filter. The observation of the angular setting is made with the telescope first on one side and then on the other side of the direct beam. The difference between the two readings is $2I_t$.

TABLE 1.1

Refractive indices of α-monobromonaphthalene for certain wavelengths of light and over a range of temperatures

Temp. (°C)	Na D (5890 Å)	Hg (orange) (5780 Å)	Hg (green) (5461 Å)	Hg (blue) (4358 Å)
15	1·6605	1·6619	1·6676	1·7057
16	1·6601	1·6615	1·6672	1·7053
17	1·6596	1·6610	1·6667	1·7048
18	1·6592	1·6606	1·6663	1·7044
19	1·6587	1·6601	1·6658	1·7039
20	1·6582	1·6596	1·6653	1·7034
21	1·6578	1·6592	1·6649	1·7030
22	1·6573	1·6587	1·6644	1·7025
23	1·6569	1·6583	1·6640	1·7021
24	1·6564	1·6578	1·6635	1·7016
25	1·6559	1·6573	1·6630	1·7011

EXAMPLE

A plate of anhydrite, $CaSO_4$, cleaved parallel to the plane (001), was used on the Kohlrausch refractometer with two orientations. The plane of the plate was vertical in both orientations but in the observations (a) and (b) the x-axis was vertical and in the observations (c) and (d) the y-axis was vertical.

(a) The polaroid between the eyepiece and the eye was orientated so as to transmit light vibrating in a vertical direction.

The difference in the readings at 20°C and for $\lambda = 5780$ Å corresponding to the boundary between the light and dark fields seen in the telescope on opposite sides of the incident beam was given by

$$2I_t = 152° \ 38'.$$

Applying the formula

$$n_c = n_l \sin I_t,$$

(n_c, n_l refer to the crystal and liquid respectively)

we obtain

$$n_c = 1{\cdot}613.$$

The corresponding vibration direction is parallel to the x-axis.

(b) The polaroid was set to transmit light vibrating in a horizontal plane.

The difference in the readings corresponding to the boundary on opposite sides of the direct beam was 141° 54′.

The corresponding refractive index, namely, 1·569, applies, therefore, to a vibration direction parallel to the z-axis.

(c) The anhydrite plate was now turned through 90° in its own plane. The polaroid was set to transmit light vibrating in a vertical plane, i.e. parallel to the y-axis.

The difference in the readings corresponding to the boundary on opposite sides of the incident beam was 143° 10′ and the corresponding refractive index 1·575. The corresponding vibration direction is parallel to the y-axis.

(d) When the polaroid was set to transmit light vibrating in a horizontal plane the same result as that given under (b) was obtained.

Thus the principal refractive indices α, β, and γ are respectively 1·569, 1·575, and 1·613. Further, the vibration directions corresponding to α, β, and γ are z, y, and x respectively. Hence, the optic sign is positive and the acute bisectrix is parallel to the x-axis.

1.6.3. *Determination of double refraction by the banded spectrum*

1.6.3.1. *Principle of the method*

On passing light from a polarizer through a plate cut from a doubly refracting crystal, light travels in the crystal as two beams plane polarized at right angles to one another. Since these beams travel with different velocities, a path-difference $t(n_1 - n_2)$ is introduced, where n_1 and n_2 are the refractive indices for this particular direction of travel, and t is the thickness of the plate as was shown in section 1.5. If this path-difference is equal to an integral number p of wavelengths the two beams on emerging from the crystal will recombine to give a beam plane polarized in the same azimuth as the incident beam. Thus any wavelength λ_0 that satisfies the equation

$$p\lambda_0 = t(n_1 - n_2) = t\Delta \tag{1.1}$$

will be extinguished if the plate be introduced between crossed Nicols or polaroids. In this equation p is an integral number, and Δ stands for $(n_1 - n_2)$. The value of $(n_1 - n_2)$ is generally dependent to a slight extent on the wavelength, but in the first place we may neglect this change. If white light is passed through the plate there will be a number of wavelengths $\lambda_0, \lambda_1, \lambda_2 \ldots$ such that

$$p\lambda_0 = (p+1)\lambda_1 = (p+2)\lambda_2. \tag{1.2}$$

For all these wavelengths the light will be extinguished. If this light is analysed into a spectrum by a grating or prism, dark bands will occur corresponding to the wavelengths $\lambda_0, \lambda_1 \ldots$, etc., in eqn. (1.2). By observations on the angle of deviation of the light corresponding to each band and by calibrating the spectrometer it is possible to obtain the values of $\lambda_0, \lambda_1 \ldots$, etc. If two bands corresponding to λ_0 and λ_n are chosen we have

$$p\lambda_0 = (p+n)\lambda_n. \tag{1.3}$$

Hence

$$p = n\lambda_n/(\lambda_0 - \lambda_n). \tag{1.4}$$

This value will not in general be integral partly because of experimental errors in determining the wavelengths and partly because of the effect of the change of the double refraction with wavelength, i.e. the dispersion of the double refraction. It we take the nearest integral value of p we obtain an approximate value of the double refraction from eqn. (1), namely,

$$\Delta = p\lambda_0/t. \tag{1.5}$$

The effect of dispersion on the values both of p and of $(n_1 - n_2)$ obtained from this measurement can be seen in the following way. We assume that the variation of the double refraction is proportional to the change of wavelength so that we may write

$$(p+n)\lambda_n = t\{\Delta + \epsilon(\lambda_0 - \lambda_n)\}, \tag{1.6}$$

where ϵ is a constant. Combining eqns. (1.6) and (1.1) we have

$$p(\lambda_0 - \lambda_n) - n\lambda_n = -t\epsilon(\lambda_0 - \lambda_n)$$

and, on reduction, we obtain

$$n = \frac{1}{\lambda_n}\lambda_0(p + t\epsilon) - (p + t\epsilon). \tag{1.7}$$

Thus, if n be plotted against $1/\lambda_n$ the slope of the line gives $\lambda_0(p + t\epsilon)$ and the intercept on the vertical axis also gives $(p + t\epsilon)$. Thus is it not possible to obtain independent values of p and ϵ from this experiment. However, if, for example, the value of Δ is known from prism spectrometer or total reflection refractometer measurements, the value of p is given by eqn. (1) and then ϵ can be obtained from the value of $(p + t\epsilon)$ derived from the graph corresponding to eqn. (1.7).

1.6.3.2. *Experimental details*

A spectrometer may be used with a grating having a known number of lines per centimetre or, if this is not available, the spectrometer readings may be calibrated by using several monochromatic wavelengths, such as are provided by the spectrum of a mercury-discharge lamp. The following details refer mainly to the setting up of the spectrometer when the former arrangement is to be used with light falling at normal incidence on the

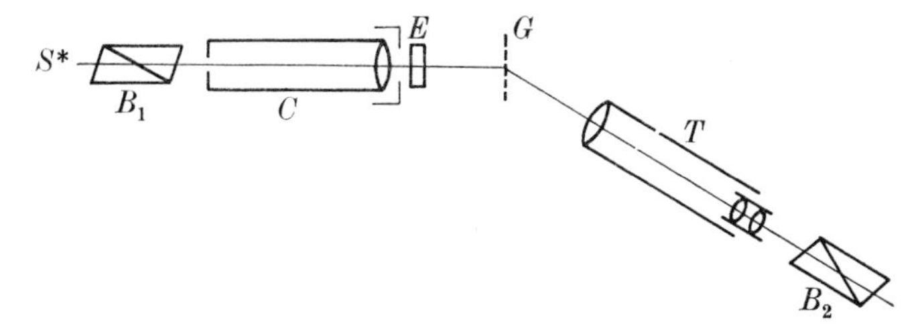

FIG. 1.7. Plan of a grating spectrometer arranged for the observation of a banded spectrum.

plane grating. The slit is illuminated by white light from the source S, Fig. 1.7, and is made plane polarized by the Nicol prism, or polaroid disk B_1. The light passes through the collimator C which is focused for infinity, and emerges as a parallel beam. The crystal plate E is mounted on the end of the collimator. Before carrying out the experiment the following preliminary adjustments must be made to the spectrometer.

1. The eyepiece of the telescope T is adjusted so that the cross wires can be clearly seen when the eye is relaxed.

2. The telescope is focused for infinity by arranging that the image of a distant object is sharp when coincident with the cross wires.

3. The collimator is made to give a beam of parallel light by viewing the image of the slit in the telescope and adjusting the collimator so that this image is sharp.

4. The grating G is mounted on the goniometer head or table with adjustments which permit it to be set with its plane perpendicular to the incident light and with its rulings parallel to the axis of rotation of the spectrometer.

5. The plane of the grating is set vertical by viewing in the telescope the image of the slit formed by reflection in the surface of the grating. When this image is symmetrical about the horizontal cross wire, the plane of the grating is vertical.

6. The rulings of the grating are set vertical. This adjustment may be carried out by viewing one of the spectra formed by the grating. By tipping the grating in its own plane this spectrum can be raised or lowered until it is symmetrical about the horizontal cross wire. The grating should be mounted on the adjustable head so that it is possible to carry out this adjustment without upsetting that made in (5).

7. Finally, the plane of the grating is set normal to the axis of the collimator. For this purpose the telescope is first set so that the direct image of the slit falls on the cross wire. The telescope is now turned through 90°. The grating is rotated until the image of the slit, formed by reflection in its surface, falls on the cross wires. The grating is now inclined at 45° to the axis of the collimator; it is turned through a further 45° to bring its plane perpendicular to the incident beam.

The polarizer B_1 is now introduced and the crystal section mounted at E. On viewing the spectrum through the analyser B_2, held before the eyepiece of T and crossed relative to B_1, the bands appear. The intensity of the spectrum is a maximum when the vibration directions of the crystal plate are inclined at 45° to those of the polarizing plates. The section should be rotated about the axis of the collimator until the maximum brightness is attained.

The cross wires of the telescope are now set on a dark band at one end of the spectrum and this band is regarded as the arbitrary zero from which the bands are counted. Observations are made over the whole range of the spectrum of the bands; the angle of the telescope setting and the number of the band, counting from the zero band, is noted in each case.

EXAMPLE

The grating spectrometer was calibrated using a mercury-vapour lamp. The data were as shown in Table 1.2.

TABLE 1.2

Calibration of spectrometer readings for known wavelengths

Light		
Colour	λ (Å)	Spectrometer reading
Red	6152	9° 20′
Orange	5780	7° 47′
Green	5461	6° 28′
Blue	4358	1° 47′

A graph was plotted and a linear relation found to exist between the wavelength and the spectrometer reading.

Using a plate of quartz cut parallel to the optic axis a banded spectrum was obtained and the scale readings given in Table 1.3 were obtained for the successive bands starting from the red end of the spectrum.

TABLE 1.3

The wavelengths corresponding to successive bands in the banded spectrum

Number of band n	Spectrometer reading	λ_n (Å)	$\lambda_0 - \lambda_n$ (Å)	p
0	9° 55′	6290	—	—
1	8° 59′	6070	220	27·6
2	8° 2′	5840	450	26·0
3	7° 19′	5638	652	25·9
4	6° 32′	5485	805	27·2
5	5° 50′	5320	970	27·4
6	5° 7′	5150	1140	27·1
7	4° 31′	5005	1285	27·2
8	3° 57′	4868	1422	27·4

The values under the heading λ_n were read off from the graph of spectrometer reading against λ_n.

The value of p is taken to be 27, the nearest integral number given by the last column of the table. The thickness t was measured and found to be 1·68, 1·67, 1·67 mm (mean 0·1673 cm).

From the approximate relation

$$\Delta = p\lambda_0/t$$

we obtain

$$\Delta = 0{\cdot}0101.$$

A graph is plotted of n against $1/\lambda_n$, from which the slope is obtained, giving

$$\lambda_0(p + t\epsilon) = 1{\cdot}70 \times 10^{-3}.$$

Thus

$$p + t\epsilon = 1{\cdot}70 \times 10^{-3}/6290 \times 10^{-8} = 27{\cdot}1.$$

If Δ be put equal to the accepted value (Kaye and Laby 1949), namely 0·0091, then

$$p = \Delta t/\lambda_0 = 0{\cdot}0091 \times 0{\cdot}1673/6290 \times 10^{-8} = 24{\cdot}2.$$

Hence

$$t\epsilon = 27{\cdot}1 - 24{\cdot}2 = 2{\cdot}9$$

and

$$\epsilon = 2{\cdot}9/0{\cdot}1673 = 17{\cdot}3.$$

The difference in the double refraction between wavelengths 6290 and 4868 Å is thus

$$\epsilon(\lambda_0 - \lambda_n) = 17{\cdot}3 \times 1422 \times 10^{-8}$$
$$= 2{\cdot}5 \times 10^{-4}.$$

We have assumed Δ for $\lambda = 6290$ Å to be 0·0091 and have calculated Δ for $\lambda = 4868$ to be 0·00935. (The value given by Kaye and Laby is 0·0093.)

It should be noted that when, according eqn. (1.2), Δ is assumed not to change with wavelength, the error in the determination of Δ is 10 per cent. The dispersion of Δ is therefore an important factor in its determination by this method.

1.6.4. *The measurement of double refraction using convergent light figures*

1.6.4.1. *Principle of the method*

When light travels in uniaxial crystals in a direction inclined to the optic axis the path difference introduced between the doubly refracted beams is given by the product of the thickness traversed and the birefringence in the direction of the light. If the ordinary and extraordinary indices of refraction are denoted n_o and n_e respectively and the refractive index in a direction inclined at an angle θ to the optic axis (within the crystal) is denoted by n, then it may be shown that

$$\tan^2\theta = \frac{n_e^2(n - n_o)(n + n_o)}{n_o^2(n_e - n)(n_e + n)}. \tag{1.8}$$

To simplify this expression we write

$$n = n_o + \delta, \qquad n_e = n_o + \Delta.$$

It must be noted that for an optically positive crystal ($n_e > n_o$) both δ and Δ are positive but for a negative crystal ($n_e < n_o$) they are both negative. When these substitutions are made in eqn. (1.8) and products of small quantities, for example δ^2, are neglected we obtain

$$\tan^2\theta = \frac{\delta}{\Delta - \delta}\cdot\left(1 + \frac{3\Delta}{2n_o}\right). \tag{1.9}$$

As θ tends to zero δ also tends to zero. For small values of θ we may write $\sin\theta$ instead of $\tan\theta$. Further, since ϕ, the measured angle, is given by

$$\sin\phi = n_o \sin\theta,$$

eqn. (1.9) further simplifies to

$$\sin^2\theta = \frac{\sin^2\phi}{n_o^2} = \frac{\delta}{\Delta}\left(1+\frac{3\Delta}{2n_o}\right). \tag{1.10}$$

It is unusual for the term in brackets to differ from unity by more than 10 per cent and to a first approximation it may be put equal to unity. When a plate of thickness t, cut perpendicular to the optic axis is used, the light inclined at an angle θ within the crystal to the optic axis, traverses a distance $t \sec\theta$ in the crystal. The condition for destructive interference between the doubly refracted beams, when using crossed polarizer and analyser, is that the path difference between the two beams is a multiple of λ. This may be expressed as

$$p\lambda = t\sec\theta(n-n_o).$$

For small θ angles this expression becomes

$$p\lambda = t\delta. \tag{1.11}$$

Combining eqns. (1.10) and (1.11) we obtain

$$\sin^2\phi = \frac{n_o^2}{\Delta}.\frac{p\lambda}{t}$$

or

$$p/\sin^2\phi = \frac{t}{n_o^2\lambda}.\Delta. \tag{1.12}$$

When all rays of a parallel beam are brought to a single point in the eyepiece field of a telescope the picture seen is known as a *directions image*. If the crystal is set with its optic axis parallel to the axis of the telescope this directions image can be seen, provided light falls upon the crystal from different directions. Then each point in the field of view of the telescope corresponds to a parallel beam passing through the crystal. The distance of the point from the centre of the field of view is approximately proportional to the inclination of the parallel beam to the optic axis. Thus under these conditions a series of circular rings is seen surrounding the dark centre that corresponds to the direction of zero double refraction. The first dark ring corresponds to a path difference of λ, the second to 2λ, etc.

If the crystal is illuminated with monochromatic light and rotated about a vertical axis in front of the telescope a number of rings will be seen to sweep across the field of view. The angle of rotation required to bring the pth ring to the intersection of the cross wires is the angle we have denoted ϕ. As may be seen from eqn. (1.12) a linear relation between p and $\sin^2\phi$ exists, at least to a first approximation.

From the counted number, p, the measured angle, ϕ, and the known values of t and n_o, the value of Δ may be found.

This method may be applied to biaxial crystals. The directions image in this case consists of two optic 'eyes' surrounded by more or less circular rings close to the points of emergence of the optic axes. Along the optic axes the birefringence is zero, at points round the first ring it is λ and for the pth ring it is $p\lambda$. At some distance from the optic eyes the curves become 'figure-of-eight' curves and thereafter any one curve encloses both optic eyes. The point midway between two optic eyes corresponds to the acute bisectrix and if there are p rings between the optic eye and this centre point the path difference between the beams travelling along the acute bisectrix is $p\lambda$. The double refraction in this case along the acute bisectrix is given by

$$p\lambda = t(\beta - \alpha) \tag{1.13}$$

for a positive biaxial crystal in which

$$(\gamma - \beta) > (\beta - \alpha).$$

For a negative biaxial crystal

$$(\gamma - \beta) < (\beta - \alpha)$$

and the equation corresponding to (1.6) becomes

$$p\lambda = t(\gamma - \beta).$$

The second birefringence, $(\gamma - \beta)$ for a positive and $(\beta - \alpha)$ for a negative crystal may be found in a manner very similar to that used for a uniaxial crystal. If we start to count the fringes from the acute bisectrix, i.e. the centre of the field, and move along a direction perpendicular to the plane containing the optic axes, then if q is the number obtained and ϕ the angle of inclination to the acute bisectrix, in air, of this qth fringe,

$$q\lambda = \frac{t \sin^2 \phi}{\beta^2}(\gamma - \beta) \tag{1.14}$$

for a positive crystal. For a negative crystal $(\beta - \alpha)$ must be substituted for $(\gamma - \beta)$.

In the example on aragonite, $CaCO_3$, given below, the double refraction is unusually high and it is necessary to take into account the term $[1 + (3\Delta/2n_o)]$ in eqn. (1.9). When used for a positive biaxial crystal eqn. (1.9) must be changed, putting

$$\Delta = \gamma - \beta \quad \text{and} \quad n_o = \beta$$

so that

$$\tan^2 \theta = \frac{\delta}{(\gamma - \beta)}\left\{1 + \frac{3(\gamma - \beta)}{2\beta}\right\}.$$

Neglecting the difference between $\tan\theta$ and $\sin\theta$ and putting $\delta = q\lambda$, we have

$$\frac{\sin^2 \phi}{\beta^2} = \frac{q\lambda}{(\gamma - \beta)}\left\{1 + \frac{3(\gamma - \beta)}{2\beta}\right\}\frac{1}{t}$$

or

$$(\gamma - \beta) = \frac{q}{\sin^2 \phi} . \beta^2 \lambda \left\{ 1 + \frac{3(\gamma - \beta)}{2\beta} \right\} \frac{1}{t} . \tag{1.15}$$

The correction term can be neglected to find a preliminary value of $(\gamma - \beta)$ and then by successive approximations a truer value can be obtained.

For a negative biaxial crystal $(\alpha - \beta)$ replaces $(\gamma - \beta)$ in the correction term, since, as noted above Δ is defined as being negative for such a crystal.

1.6.4.2. *Experimental arrangement*

Monochromatic light is used by inserting an appropriate filter between the mercury lamp and the collimator lens. No slit or limiting aperture is required in the collimator and a 32-mm focal length lens gives a convenient convergence to the incident light. A polarizing filter is also placed between the lamp and the collimator lens. (The polarizer and the colour filter fit into one another and into the cup on the end of the collimator tube.) The crystal is mounted on the goniometer head. The analyser is placed between the crystal and the 3-in focal length lens that is used in the telescope in place of the double lens normally used on an optical goniometer. The reason for this change is that in many lenses there is sufficient mechanical strain so that when viewed between crossed polaroids they show strain figures of their own. These are confusing and can be eliminated by placing the analyser over the telescope objective instead of over the eyepiece. To make this possible the double lens must be replaced by a single lens. During the observations the telescope remains fixed and only the crystal is rotated.

1.6.4.3. *Experimental procedure*

Care should be taken to ensure that the telescope is properly focused so that a distant object is clearly seen without parallax in the eyepiece field. The vibration directions of the polarizer and analyser are set at 45° to the horizontal plane and crossed with respect to one another. The vibration direction of a polarizing filter can be found by looking at the reflection from a polished surface at a glancing angle of about 35°. As the polarizer is rotated the intensity of the light transmitted rises and falls. When it is a minimum the vibration direction of the polarizer lies in the plane containing the incident and reflected rays. In this way the vibration direction transmitted by the polarizer may be found.

The crystal in the form of a plate with parallel major faces is mounted on the goniometer head. If it is uniaxial there is no other adjustment required and the angles ϕ corresponding to the successive rings can be read off on the scale attached to the vertical shaft carrying the crystal.

If a biaxial crystal is used it must first be set with its optic axes lying in a horizontal plane. To achieve this it may be necessary to rotate the crystal plate in its own plane, and to examine the ring system in each setting. In every case the acute bisectrix, which will usually be the normal to the plane of the plate, must be perpendicular to the axis of rotation of the crystal. The apparent angle between the two optic axes, in air, denoted $2E$, can be measured and also p the number of rings between an optic eye and the centre of the field.

The crystal is then turned through 90° in its own plane and the number of rings q between the centre and a certain direction defined by the angle ϕ is observed. From these observations the values of the two double refractions and the optic axial angle can be found.

EXAMPLES

(a) A plate of quartz cut perpendicular to the optic axis and 0·296 cm thick was used. Observations were made in mercury orange light ($\lambda = 5780$ Å). The rings were not as sharp as the usual images obtained by reflections from crystal faces and to obtain an estimate of the accuracy with which the observations could be repeated two consecutive sets of observations were made and are given in Tables 1.4 and 1.5.

TABLE 1.4

The variation of the angle 2ϕ for successive rings in the uniaxial optic picture

	Inclination of fringes to optic axis					
Number of ring	Crystal setting			Crystal setting		
p	Left	Right	2ϕ	Left	Right	2ϕ
1	4° 0′	339° 14′	24° 46′	4° 21′	338° 53′	25° 28′
2	9° 59′	333° 15′	36° 44′	9° 59′	333° 15′	36° 44′
3	14° 19′	328° 54′	45° 25′	14° 19′	328° 49′	45° 30′
4	17° 59′	325° 11′	52° 48′	17° 54′	325° 12′	52° 42′
5	21° 4′	321° 53′	59° 11′	21° 15′	321° 57′	59° 18′
6	24° 9′	318° 55′	65° 14′	24° 6′	318° 58′	65° 8′

TABLE 1.5

Calculation of $p/\sin^2\phi$ for successive rings

Number of ring p	ϕ mean	$\sin^2\phi$	$p/\sin^2\phi$
1	12° 34′	0·0475	21·1
2	18° 22′	0·0992	20·1
3	22° 44′	0·1490	20·1
4	26° 23′	0·1971	20·3
5	29° 38′	0·2440	20·5
6	32° 36′	0·2905	20·7

The value of $p/\sin^2\phi$ that we require is that obtained when ϕ is zero. To obtain this value $p/\sin^2\phi$ is plotted against ϕ and the straight line through the points corresponding to rings 3–6 is extrapolated to $p = 0$. The value so obtained is 19·8.

The mean refractive index n is taken as 1·55. Hence, from the eqn. (1.12),

$$p\lambda = \frac{t\sin^2\phi}{n^2}.\Delta,$$

we obtain

$$\Delta = \frac{p}{\sin^2\phi}.\frac{\lambda n^2}{t} = 19{\cdot}8 \times \frac{5780 \times 10^{-8} \times 1{\cdot}55^2}{0{\cdot}296} = 0{\cdot}0093.$$

(Kaye and Laby, *Physical and Chemical Constants*: $\Delta = 0{\cdot}0091$.)

(b) A crystal of aragonite, $CaCO_3$, cut perpendicular to the acute bisectrix, was examined in mercury orange light ($\lambda = 5780$ Å). The thickness of the crystal was 0·198 cm. The crystal is optically negative so that for light travelling along the acute bisectrix the double refraction is $(\gamma - \beta)$.

Determination of $(\gamma - \beta)$

The number of rings between the centre of the optic picture and an optic eye was 15¼. (There is difficulty in estimating a fraction of a fringe.) In eqn. (1.5), $t = 0{\cdot}198$ cm, hence

$$\gamma - \beta = 15{\cdot}25 \times 5780 \times 10^{-8}/0{\cdot}198 = 0{\cdot}0045.$$

Determination of $(\beta - \alpha)$

The crystal plate was mounted on a single-circle goniometer with its optic axial plane vertical and observations on the angular settings corresponding to the successive fringes were made on both sides of the centre of the optic picture. The observations were repeated and the mean value of ϕ recorded in Table 1.6.

TABLE 1.6

Values of $q/\sin^2\phi$ for successive fringes in plane normal to optic axial plane

Number of fringe q	ϕ mean	$\sin^2\phi$	$q/\sin^2\phi$
¾	3° 31′	0·00377	199
1¾	5° 20′	0·00865	203
2¾	6° 39′	0·01341	205
3¾	7° 43′	0·01804	208
4¾	8° 42′	0·02289	208
5¾	9° 34′	0·02763	208
6¾	10° 20′	0·03218	210
7¾	11° 3′	0·03674	211
8¾	11° 44′	0·04135	212
9¾	12° 23′	0·04599	212

When extrapolated to $\phi = 0$ the straight line passing through the points $3\frac{3}{4}$–$9\frac{3}{4}$ gives a value of $q/\sin^2\phi$ equal to 204. The points $\frac{3}{4}$–$2\frac{3}{4}$ can be neglected because of the difficulty of estimating $\frac{3}{4}$. The mean index of refraction is 1·681 and hence, (eqn. 1.14)

$$\beta - \alpha = 204 \times 5780 \times 10\ ^{8} \times 1{\cdot}681^2/0{\cdot}198$$

$$= 0{\cdot}168.$$

The correction term given in eqn. (1.9) is in this case

$$\left(1 + \frac{3(\alpha - \beta)}{2\beta}\right)$$

and the value of this using the first approximation to the value of $(\beta - \alpha)$ is 0·85. This gives a corrected value of $(\beta - \alpha)$ of 0·143. Thus we obtain finally,

$$\alpha = 1{\cdot}538, \qquad \beta = 1{\cdot}681, \qquad \gamma = 1{\cdot}686.$$

(Phillips, A. H., *Mineralogy*: $\alpha = 1{\cdot}529$, $\beta = 1{\cdot}681$, $\gamma = 1{\cdot}685$.)

1.6.5. *The study of rotatory polarization*

1.6.5.1. *Fresnel's experiment*

Fresnel demonstrated that in an optically active uniaxial crystal light travels along the optic axis in the form of two circularly polarized beams with different refractive indices and opposite senses of rotation. This may be shown with the following apparatus. It consists of a telescope T (Fig. 1.8) and a collimator in line with it, C, between which is placed a triple

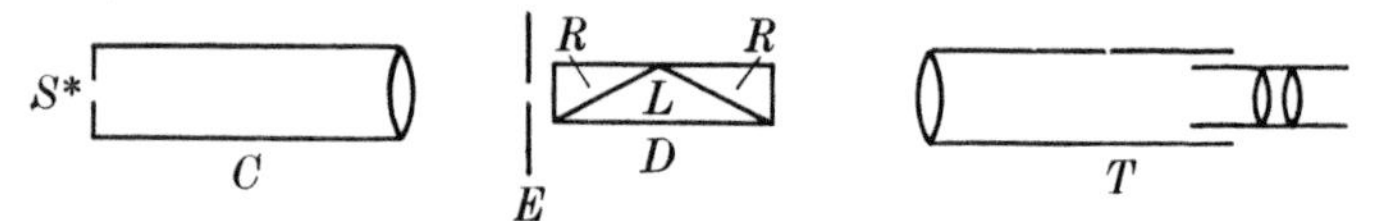

FIG. 1.8. Diagram of Fresnel's apparatus for showing the nature of rotatory polarization.

quartz prism D. This prism consists of two pieces of one hand of rotation (e.g. right) and a third piece of the opposite hand. In all three the optic axis is accurately parallel to the length of the final prism and the pieces are cemented together with Canada balsam. A diaphragm E only permits light to reach the telescope after passing through the triple prism. The collimator C has a very narrow slit illuminated by a source S. Before the prism is put into place the image of the slit seen in the telescope is single; after inserting the prism it is a double image. Although unpolarized light is passed into the prism each of the two images is circularly polarized but the one has the opposite sense of rotation to the other. This can be shown by placing a quarter-wave plate between D and T and using a polaroid disk or Nicol prism between T and the observer. For this observation monochromatic light must be used because a quarter-wave plate is of a thickness appropriate to only one colour, usually yellow. On rotating the analyser the images are seen to extinguish one after the other, the vibration directions being mutually perpendicular. Since a quarter-wave plate renders circularly polarized light plane polarized and the direction of vibration of the plane-polarized beam depends on the sense of rotation of the circularly polarized beam, it is clear that this demonstration proves the truth of Fresnel's hypothesis concerning the nature of the vibrations when light travels along the optic axis of a uniaxial crystal.

1.6.5.2. *The measurement of specific rotatory polarization*

Owing to the fact that a right-handed circularly polarized beam travels faster (or slower) than a left-handed circularly polarized beam along the optic axis of a uniaxial crystal, which is optically active, the resultant plane of vibration of the emergent plane-polarized light is rotated with respect to the plane of vibration of the incident light. The amount of this rotation for unit thickness of the crystal (and for a given wavelength of light) is known as the specific rotation of the crystal. This may be determined for basal sections of quartz of a few millimetres thickness using a modified microscope. The condenser, objective, and eyepiece are first removed. An analyser, either a Nicol prism or a piece of polaroid, is mounted on a small bridge on the rotating stage of the microscope. The angle of rotation of the stage must be capable of being read on a divided scale. The quartz plate is mounted on the stage under the analyser. It makes no difference whether the quartz rotates with it or is stationary. With the polarizer, quartz plate, and analyser in place and using monochromatic light, the stage is rotated until extinction is reached. The specimen is removed and the stage rotated until extinction is again obtained. The difference between the two readings gives the rotation of the plane of polarized light due to the specimen. There is a certain ambiguity in a single observation of this kind owing to the fact that the specimen may be either right-handed or left-handed. If the rotation were $\theta°$ for one sense of rotation it would be $(180 - \theta)°$ for the other. For thick specimens there is a further ambiguity in that the rotation may be $\theta + n.\ 180°$, where n is a whole number. To eliminate these sources of ambiguity in determining the specific rotation it is necessary to use several wavelengths of light and several thicknesses of crystal plate. For all optically active crystals the rotation increases linearly with the thickness and unless the material is coloured the specific rotation increases as the violet end of the spectrum is approached. When the same value for the specific rotation of several specimens of different thickness in one colour of light has been obtained, it is certain that the ambiguities have been overcome.

The accuracy with which the angle of rotation can be determined depends on the sharpness with which the extinction can be observed. If the crystal is imperfect or the surfaces are not properly polished it may be difficult to determine the extinction position correct to 1°. If the crystal is not cut perpendicular to the optic axis within one or two degrees it may be impossible to obtain a correct value for the specific rotation. Finally, if the light is not truly monochromatic, difficulty may be experienced in obtaining a proper extinction. If none of these sources of error is present the thicker the specimen in the direction of observation the greater the angle of rotation and the greater the accuracy in the determination of the specific rotation.

EXAMPLE

Using plates of quartz and a mercury-vapour lamp with colour filters to separate the yellow, green, and blue lines, the results shown in Table 1.7 were obtained for the angular settings of the polarizer. All three specimens produced an anticlockwise rotation of the plane of polarization and were therefore left-handed (denoted L).

TABLE 1.7

Relation between the angle of rotation, the thickness of crystal plate, and the colour of light

Thick-ness (mm)	Crossed polaroids settings	Colour of light					
		Yellow		Green		Blue	
		Setting	Rotation	Setting	Rotation	Setting	Rotation
1·07	104°	81°	23°L	78°	26°L	62°	42°L
		81°	23°	79°	25°	61°	43°
2·63	105°	46°	59°L	37°	68°L	356°	109°L
		46°	59°	37°	68°	355°	110°
5·98	104°	330°	134°L	308°	156°L	232°	232°L
		330°	134°	313°	151°	228°	236°
				309°	155°	218°	246°
				311°	153°	221°	243°

From the data in Table 1.7 the specific rotations given in Table 1.8 were obtained by dividing the rotation by the thickness.

TABLE 1.8

Relation between the specific rotation and the colour of the light

Thickness (mm)	Colour of light		
	Yellow	Green	Blue
1·07	21°·5	23°·8	39°·7
2·63	22°·4	25°·9	41°·6
5·98	22°·4	25°·8	40°·0
Mean value	22°·1	25°·2	40°·4

1.6.6. *Determination of optical constants of transparent and opaque substances by using reflected light*

When light is incident upon a plane surface of either a transparent or an opaque substance, the state of polarization of the reflected beam may be determined in terms of the amplitudes and phase relations of the components polarized respectively parallel and perpendicular to the plane containing the incident and reflected beams. The intensity of reflection of a plane-polarized beam depends upon the inclination of the plane of

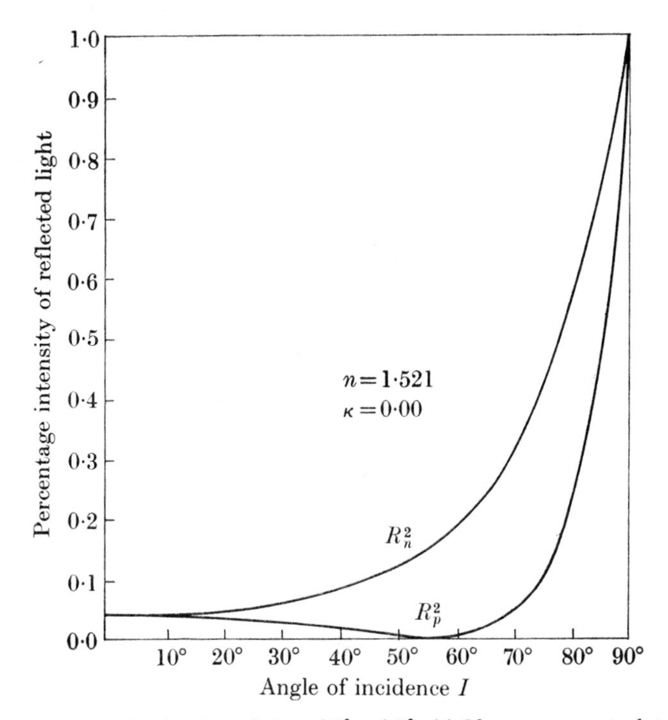

FIG. 1.9. Graphs showing the variation of R_n^2 and R_p^2 with I for a transparent substance.

vibration to the plane of incidence. If the reflected intensities R_p^2 and R_n^2 (denoting respectively the intensities of the components parallel and perpendicular to the plane containing the incident and reflected beams) are plotted against I (the angle between the incident or reflected beam and the normal to the reflecting surface), we obtain two curves. The curve for R_p^2 descends to zero for a particular value of I (Fig. 1.9) called the *Brewster angle* or *principal angle of incidence*. The tangent of this angle is equal to the refractive index of the medium.

For an opaque reflecting surface the corresponding curves showing the relation between R_n^2, R_p^2, and I have the general form shown in Fig. 1.10.

There are two features of the curves such as those shown in Figs. 1.9 and 1.10 which are commonly utilized, (a) the intensity of reflection ($R_p^2 = R_n^2$) at an angle $I = 0$, and (b) the value of I for which R_p is zero (for transparent substances only).

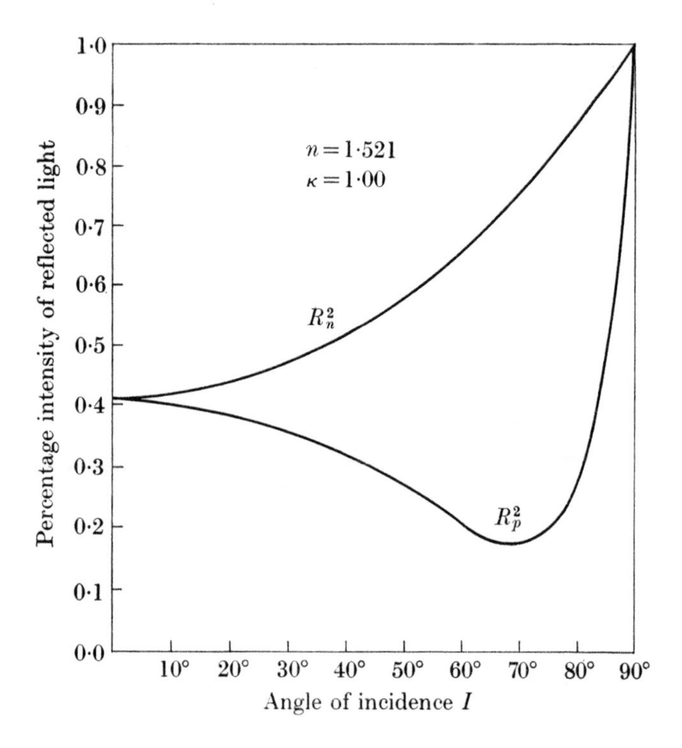

FIG. 1.10. Graphs showing the variation of R_n^2 and R_p^2 with I for an opaque substance.

1.6.6.1. *Determination of refractive indices and absorption coefficients of opaque crystals*

Measurements at normal incidence can never yield more than the reflecting powers at normal incidence and the variation of these with plane of polarization of the incident light and with colour. These quantities may be enough for the purpose in hand but more fundamental quantities are the refractive index and absorption coefficient. In an opaque medium the term refractive index cannot have the same meaning as it does in a transparent medium since there is no observable refracted light. However, the generalization of the electromagnetic theory of light, which has proved adequate to correlate all the observed phenomena, involves an imaginary refracted beam making an angle r' with the normal to the surface. If I is the angle of incidence and n' the imaginary refractive index, then Snell's law applies, i.e.

$$n' = \frac{\sin I}{\sin r'}.$$

This imaginary refractive index may be resolved into a real and an imaginary part as is shown in the equation

$$n' = n - ik, \qquad \text{where } i = \sqrt{(-1)}.$$

The real part n is the ordinary refractive index, as may be seen by putting the absorption coefficient k equal to zero. The absorption coefficient is defined by the diminution of the intensity of a beam of light in passing through a thickness t of the specimen. If I_t and I_0 are the intensities of the transmitted and incident beams respectively, then

$$I_t = I_0 e^{-(4\pi/\lambda)kt}.$$

It should be noted that when k is unity, t must be of the order of 10^{-5} cm if the ratio of I_t/I_0 is not to be less than 0·1. It is therefore usually impossible to obtain specimens thin enough to measure k directly by finding the ratio I_t/I_0.

1.6.6.2. *Change of phase on reflection*

It is possible to determine n and k by tracing with a spectrophotometer the curves such as those given in Fig. 1.10 for the relation between the intensities of reflection of the parallel and perpendicular components and the angle of incidence. The formulae for this method are clumsy and the evaluation lengthy, so an alternative method is normally employed. This depends on finding the changes of phase, which both components experience on reflection, as the angle I is varied. At normal incidence both components parallel and perpendicular to the plane of incidence become indistinguishable since the plane of incidence may have any azimuth. It is well known from studies on Newton's rings and other interference experiments that there is a change of phase of π on reflection at normal incidence from a medium of higher index. This change is experienced by both components when $I = 0$. When $I = 90°$ and the light just grazes the reflecting surface the perpendicular component still has an electric vibration which is the same as at normal incidence relative to the reflecting surface, i.e. it lies in the surface. For the perpendicular component there is thus a change of phase of π at $I = 0°$ and $90°$ and, for the same reason, at all other angles. For the parallel component, however, the electric vector lies in the surface when $I = 0°$ but is perpendicular to the reflecting surface when $I = 90°$. When the electric vector of the light wave is perpendicular to the reflecting surface it can induce a free charge on the surface and there is therefore no change of phase. At the Brewster angle there is an abrupt change of phase from π to 0. The variation of the changes of phase ϕ_n and ϕ_p are shown in Fig. 1.11 for transparent substances. The corresponding changes in phase for

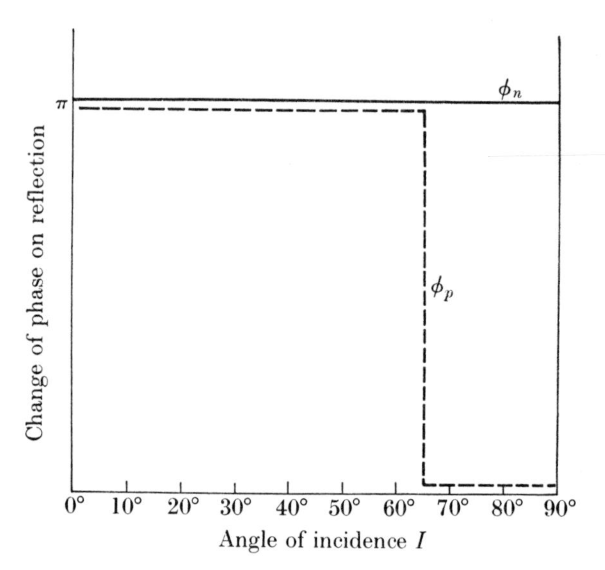

FIG. 1.11. Graphs showing for a transparent substance the changes of phase on reflection for (a) the perpendicular component ϕ_n (full line), and (b) the parallel component ϕ_p (dotted line), as a function of I.

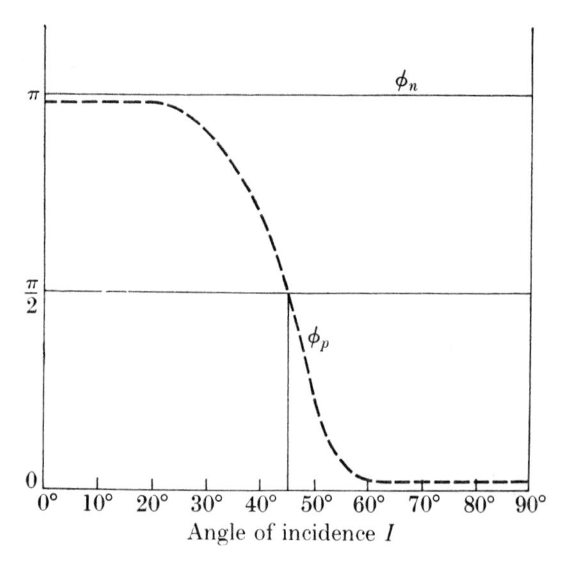

FIG. 1.12. Graphs showing for an opaque substance the changes of phase on reflection for (a) the perpendicular component ϕ_n (full line), and (b) the parallel component ϕ_p (dotted line), as a function of I.

opaque substances are illustrated in Fig. 1.12 where it will be seen that there is no longer an abrupt change of phase but a gradual one, more or less extended according to the values of n and k. The angle of incidence at which this change of phase is $\frac{1}{2}\pi$ is denoted $\bar{I}$ and called the *principal angle of incidence*. In transparent materials this angle is the same as the Brewster angle.

1.6.6.3. *Direct measurement of refractive index and absorption coefficient*

A spectrometer is arranged as shown diagrammatically in Fig. 1.13. Monochromatic light from a source S enters the slit of the collimator C

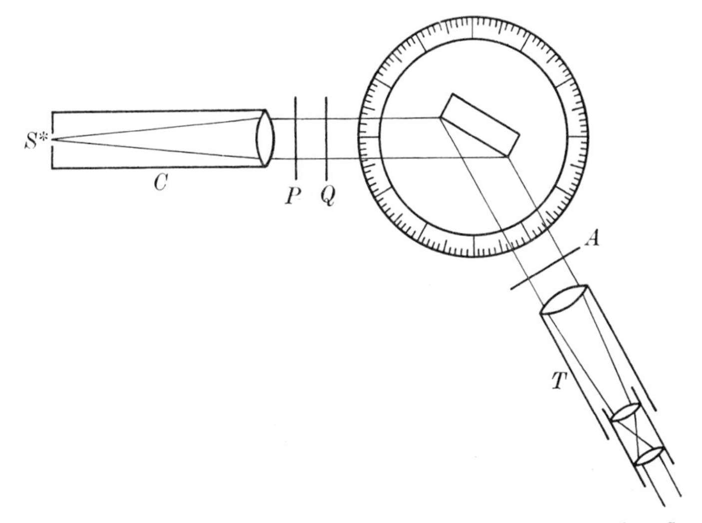

FIG. 1.13. Diagram of a spectrometer arranged for making observations by reflected light.

and is polarized by passage through a polaroid or Nicol prism P. A quarter-wave plate Q may be inserted in the path of the beam. Both P and Q may be rotated in their own planes about the light beam as axis and the position of P is given on a scale graduated in degrees. Q is not provided with a graduated scale. The crystal is mounted on a goniometer head the rotation of which about a vertical axis is given on a graduated scale. After reflection at the surface of the crystal the light passes through an analyser A and enters the telescope T.

Experimental details

To determine the principal angle of incidence the following procedure is adopted.

1. With the crystal removed T is brought into line with C. Q is removed from the light path.

2. The analyser A is now crossed with respect to P, both being in any arbitrary orientation about the axis of the telescope.
3. Q is inserted, when light will in general be transmitted through A owing to the double refraction of the quarter-wave plate. Q is now rotated to an extinction position.
4. P is rotated through 45° so that the light emerging from Q must be circularly polarized. This should be tested by rotating A right round in its own plane. No variation in the intensity of the transmitted light should be observed.
5. The crystal is set up so that its reflecting surface gives an image of the slit in the telescope which is symmetrical with respect to the intersection of the cross wires.
6. A is now rotated to reduce the intensity of the transmitted light to a minimum.
7. The crystal and the telescope are now rotated together, so as to keep the reflected image continually in the field of view, until the intensity of the image is a minimum.
8. A is readjusted to make the intensity as small as possible.
9. Keeping T stationary, the crystal is rotated so as to cause the image to sweep across the field of view. When T and A are properly set the image should vanish as it approaches the intersection of the cross wires from either side.

The angle between the direct beam and the axis of the telescope, which is $180 - 2\bar{I}$, is read off from the graduated circle of the spectrometer. Thus $\bar{I}$ is found. The angular setting of A is read. In general, the vibration direction of A will be inclined to the horizontal at an angle $\bar{\psi}$ which is only zero for transparent reflectors. As may be seen from Fig. 1.10, the intensity of the parallel component is not reduced to zero at the angle of incidence $\bar{I}$. Thus the circularly polarized beam emerging from Q is rendered plane polarized by the change of phase of $\frac{1}{2}\pi$ between the vibrations of the two components introduced on reflection. For a transparent substance the perpendicular component only is reflected and the plane-polarized beam vibrates vertically. To produce extinction the vibration direction of A is made horizontal. With an opaque substance, however, the perpendicular component has a certain small value relative to the parallel component (Fig. 1.14) and since their phase relation has been changed from $\frac{1}{2}\pi$ to either π or 0 by reflection (depending on the sense of rotation of the incident circularly polarized beam), we obtain a resultant plane-polarized beam inclined at an angle $\bar{\psi}$ to the vertical. This angle $\bar{\psi}$ is known as the *principal angle of azimuth* and is closely dependent on the absorption coefficient.

It is usual to measure $2\bar{\psi}$ rather than $\bar{\psi}$. This can readily be done, if, after noting the original setting of A, the polarizer P is turned through 90° leaving Q as before. This has the effect of reversing the sense of rotation of the circularly polarized beam incident on the crystal. In Fig. 1.14 R_p would now be drawn to the left since there would be a change of phase of π on reversing the hand of rotation of the incident beam. The new setting of A giving extinction differs from the first setting by $2\bar{\psi}$.

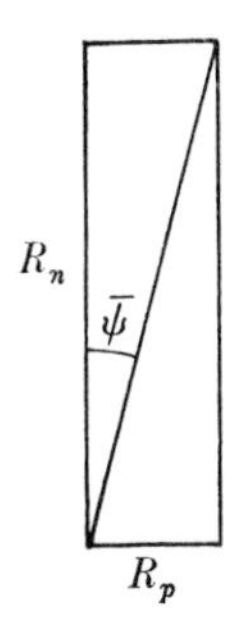

FIG. 1.14. Diagram showing the relation between R_n, R_p, and $\bar{\psi}$.

The formulae for n and k follow from the electromagnetic theory of light and are given in terms of $\bar{I}$ and $\bar{\psi}$ as follows:

$$n^2 = \tfrac{1}{2}\tan^2 \bar{I}\{1 - 2\sin^2 \bar{I}\sin^2 2\bar{\psi} + \sqrt{(1 - \sin^2 2\bar{I}\sin^2 2\bar{\psi})}\}$$

$$k^2 = \frac{\sqrt{(1 - \sin^2 2\bar{I}\sin^2 2\bar{\psi})} - (1 - 2\sin^2 \bar{I}\sin^2 2\bar{\psi})}{\sqrt{(1 - \sin^2 2\bar{I}\sin^2 2\bar{\psi})} + (1 - 2\sin^2 \bar{I}\sin^2 2\bar{\psi})}$$

It should be noted than when $\bar{\psi}$ is small so that terms in $\sin^2 2\bar{\psi}$ are small compared with unity

$$k = \sin 2\bar{\psi}\sin^2 \bar{I},$$

and when $k = 0$,

$$n = \tan \bar{I}.$$

The accuracy that can be obtained by this method depends almost entirely on the flatness of the surface and its polish. Without an excellent surface the results may be erroneous. However, with adequate polishing results correct to a few per cent can usually be obtained.

EXAMPLE

The following measurements were made on a spectrometer by eye, using the quarter-wave plate described above.

Pyrite, FeS_2: $2(90° - \bar{I}) = 26°\ 24'$, $2\bar{\psi} = 44°$.

Hence, applying the above extended formula we obtain

$$n^2 = \tfrac{1}{2}\tan^2 76°\ 48'\{1 - 2\sin^2 76°\ 48'\sin^2 44° + \sqrt{(1 - \sin^2 153°\ 36'\sin^2 44°}\},$$

$$n = 3{\cdot}07,$$

$$k^2 = \frac{\sqrt{(1 - \sin^2 153°\,36'\sin^2 44°)} - (1 - 2\sin^2 76°\,48'\sin^2 44°)}{\sqrt{(1 - \sin^2 153°\,36'\sin^2 44°)} + (1 - 2\sin^2 76°\,48'\sin^2 44°)},$$

$$k = 0{\cdot}91.$$

2. Diamagnetic and Paramagnetic Properties of Crystals

2.1. The constants that define second-order tensor properties

In the next three chapters we shall be concerned with the measurement of a group of properties known as second-order tensor properties,† which include dia- and para-magnetic properties, thermal properties, and plastic deformation. The variation of any such property can be represented by a sphere in the cubic system, an ellipsoid of revolution in the tetragonal, hexagonal, and rhombohedral systems, and a triaxial ellipsoid in the orthorhombic, monoclinic, and anorthic systems. The principal axes of these ellipsoids are parallel to the directions with which the principal coefficients are associated. These coefficients express the relations between pairs of vector quantities, for example induced magnetic moment and applied magnetic field, or heat flow and temperature gradient. Because of the relation to the crystal symmetry there is only one coefficient for each property in the cubic system, two in the tetragonal, hexagonal, and rhombohedral systems (one of which is associated with the unique axis and the other with all directions at right angles to it). In the remaining systems there are three coefficients. In the orthorhombic system the principal coefficients are associated with the crystallographic axes. In the monoclinic system one principal coefficient is associated with the axis of symmetry or the normal to the plane of symmetry, whichever is appropriate. The other two principal axes are perpendicular to the first and to each other. In the anorthic system the principal axes are always mutually perpendicular, but their direction is not limited in any way by the crystallographic axes. The coefficients of different second-order tensors are entirely independent of each other, and the ellipsoid that describes any second-order property may vary its shape and direction, within its symmetry relation, with temperature and pressure. For all second-order properties, therefore, the number of coefficients to be found depends only on the symmetry and may be specified as in the Table 2.1.

In measuring the coefficient of a second-order property it is usual to cut the crystal in such a way that one of the principal coefficients is measured directly in each experiment. This is not always possible in the monoclinic

† All the tensor statements in this book are based on Voigt's treatment of 1910. It is not yet certain that a serious revision is necessary along the lines suggested by Laval, J., *L'état solide (rapports et discussions)*, p. 273. Congress Solway Bruxelles, Stoops (1951).

TABLE 2.1

The number of physical constants required in the seven crystallographic systems

System	Number of physical coefficients to be found	Number of angles required to define the principal axes
Cubic	1	—
Tetragonal, Rhombohedral, Hexagonal	2	—
Orthorhombic	3	—
Monoclinic	3	1
Anorthic	3	3

and anorthic systems, because it may be impossible to know in advance the orientation of the principal axes of the ellipsoid representing the property. In such a crystal the method used is either one of trial and error or one based on the geometry of the ellipsoid.

2.2. The measurement of the difference between two principal magnetic susceptibilities by using the torque exerted in a uniform magnetic field

2.2.1. *Principle of the method—method A*

Most solids are diamagnetic, i.e. they are repelled in a non-uniform magnetic field from regions of strong to regions of weak magnetic field strength. In a uniform magnetic field, however, cubic crystals are unaffected by the field, but anisotropic crystals tend to orientate themselves with the least diamagnetic direction in the crystal parallel to the lines of force. These forces are all very small and require delicate apparatus to make them observable. If a crystal is suspended on a fine fibre so that it can oscillate about a vertical axis the crystal tends to set in a particular direction relative to the magnetic field. The couple exerted on the crystal by the magnetic field can be found as follows. Let us suppose that the principal susceptibilities χ_1, χ_2 in the horizontal plane are inclined at angles of θ and $(\pi/2)-\theta$ to the magnetic field. In a field of strength H the magnetic moment induced along χ_1 is due to the component $H\cos\theta$ of the field. The susceptibility is defined as the magnetic moment per unit volume induced by unit magnetic field strength. Thus the magnetic moment M_1, directed parallel to χ_1, is given by

$$M_1 = vH\cos\theta\,\chi_1$$

where v is the volume of the crystal. The component of the field perpendicular to χ_1 is $H\sin\theta$ and the couple this exerts on M_1 is $M_1 H\sin\theta$.

Now, putting in the value of M_1, we have

$$M_1 H\sin\theta = \tfrac{1}{2}vH^2\sin 2\theta\,\chi_1.$$

The component $H\sin\theta$ of the field produces a magnetic moment M_2 directed along χ_2. The value of M_2 is given by

$$M_2 = vH\sin\theta\,\chi_2.$$

The component $H\cos\theta$ of the magnetic field exerts a couple on this induced moment of $M_2 H\cos\theta$. We have

$$M_2 H\cos\theta = \tfrac{1}{2}vH^2\sin 2\theta\,\chi_2.$$

These two couples are oppositely directed so that the resultant couple G is given by

$$G = \tfrac{1}{2}vH^2\sin 2\theta(\chi_1 - \chi_2).$$

There is thus a maximum value, G_{max}, when the principal axes of susceptibility are inclined to the lines of force at an angle of 45°. We may write

$$G_{max} = \tfrac{1}{2}vH^2(\chi_1 - \chi_2).$$

For small values of θ, G is proportional to θ and the restoring couple per unit angular displacement is given by

$$\frac{G}{\theta} = vH^2(\chi_1 - \chi_2). \tag{2.1}$$

The time of oscillation of the crystal, T_0, in the absence of a magnetic field is given by

$$T_0 = 2\pi\sqrt{\frac{I}{k}}, \tag{2.2}$$

where I is the moment of inertia about the axis of oscillation and k is the restoring couple per unit angular displacement due to the suspending fibre. The value of I for a rectangular block is given by

$$I = m\frac{a^2 + b^2}{12},$$

where m is the mass and a and b are the lengths of the sides perpendicular to the axis of oscillation.

If T_1 is the time of oscillation in the presence of the magnetic field

$$T_1 = 2\pi\sqrt{\left(\frac{I}{k + G/\theta}\right)} \tag{2.3}$$

since the restoring couple due to the magnetic field is added to that of the fibre.

Combining eqns (2.1), (2.2), and (2.3) we obtain

$$\chi_1 - \chi_2 = \frac{1}{vH^2}\cdot 4\pi^2 I\left(\frac{1}{T_1^2} - \frac{1}{T_0^2}\right). \tag{2.4}$$

Thus measurements of T_1, T_0, H and the dimensions of the crystal make it possible to find the differences in the principal susceptibilities.

2.2.2. *Method B*

An alternative method depends on determining G_{max}. The crystal is supported on a rotatable graduated torsion head. This head is turned until

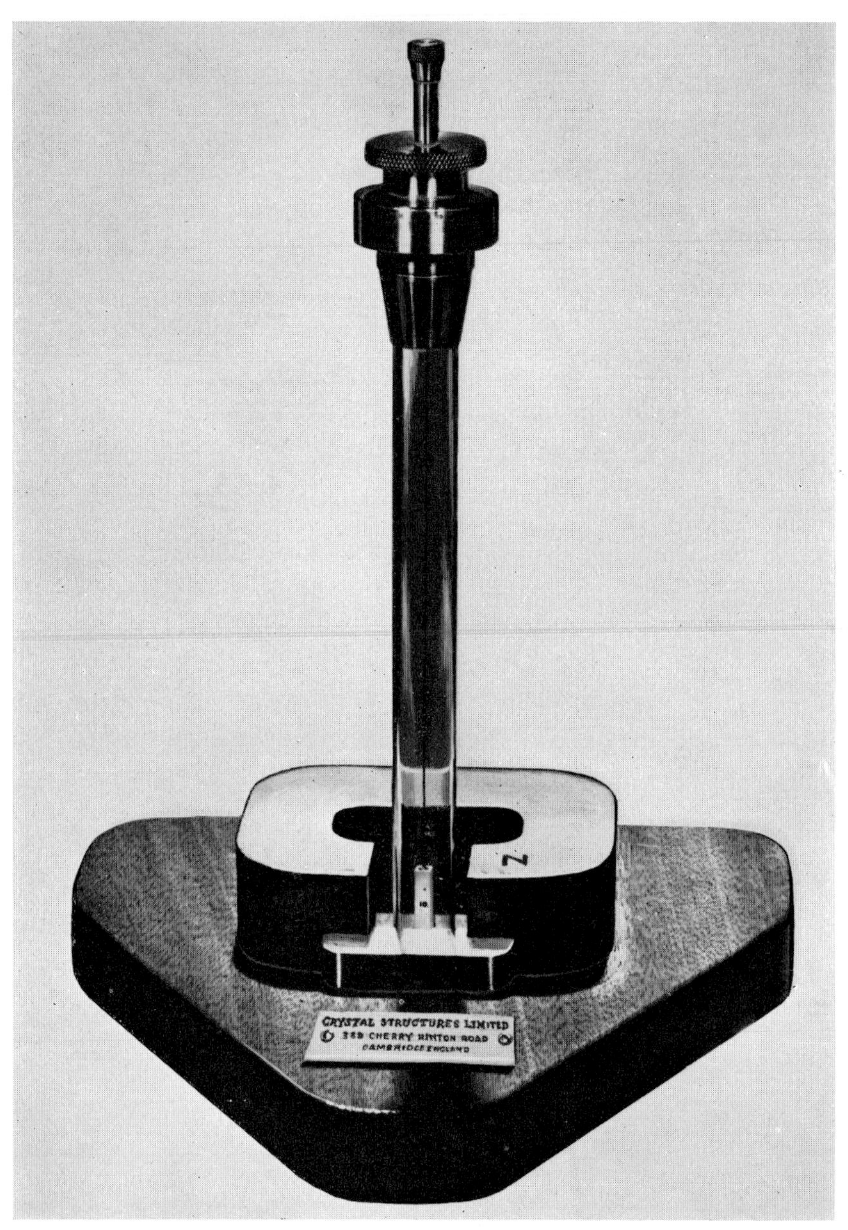

FIG. 2.1. Apparatus supporting a crystal on a fibre in the uniform field between the poles of a permanent magnet.

on putting the permanent magnet in place on the base of the instrument there is no net rotation of the crystal after it has come to rest. The torsion head is now rotated in one sense, say clockwise. After a small rotation of the torsion head the crystal is twisted through a small angle but the rotation of the crystal is much less than that of the torsion head. As the fibre is progressively twisted in the same sense the crystal is more and more rotated from its initial setting until when this rotation is 45°, or very near to this value, it becomes unstable and rotates rapidly and several times. The setting of the torsion head at which this unstable position is reached corresponds to a rotation denoted ϕ. The crystal becomes unstable because the magnetic twist decreases more rapidly than the mechanical twist when θ exceeds 45°.

Now we have

$$G_{\max} = k\left(\phi - \frac{\pi}{4}\right)$$

since the twist of the top relative to the bottom of the fibre is

$$\left(\phi - \frac{\pi}{4}\right).$$

Also,

$$G_{\max} = \tfrac{1}{2}vH^2(\chi_1 - \chi_2).$$

Hence

$$\chi_1 - \chi_2 = \frac{2k}{vH^2}\left(\phi - \frac{\pi}{4}\right).$$

The value of k can be found from T_0, using eqn (2.2).

The experiment should be repeated with the opposite sense of rotation of the torsion head, i.e. anti-clockwise. The mean of the two values of ϕ eliminates any error in the initial setting.

2.2.3. *Experimental arrangement*

The crystal provided is a rectangular block of calcite cut with its optic axis parallel to one of the shorter sides. When the crystal is suspended by the fine suspension strip between the poles of the permanent magnet the optic axis tends to set perpendicular to the lines of force (Fig. 2.1). The suspension strip is attached to a torsion head graduated in degrees which is supported by a glass tube mounted on a wooden base. There is on the base a brass stop against which the magnet should be placed when the crystal is to be subjected to the magnetic field. A stop watch or other timing device completes the apparatus required for this experiment. (Care should be taken to keep the watch away from the magnet.)

2.2.4. *Conduct of the experiment*

Method A

The torsion head should be turned so that when the magnet is gently placed on the base board against its brass stop, the crystal does not rotate from its previous resting position. The setting of the torsion head is noted. The torsion head is now rotated through a certain angle and then returned to its initial setting. This sets the crystal oscillating and its time of oscillation is found both in and out of the magnetic field. The strength of the magnetic field must be found by one or other of the standard methods, for example using a search coil and a fluxmeter. The mass of the crystal is given and so are the linear dimensions. (It is easy to break the suspension and it is not recommended that the crystal should be removed from the glass tube for weighing and measuring.)

Method B

The initial setting is made as in method A and the corresponding reading of the torsion head is taken as the zero reading. The experiment should be performed in two stages. At first the torsion head is turned, not very slowly, until the unstable position is reached and the approximate setting of the graduated head is noted. The torsion head and crystal are returned to their initial settings again and the twisting of the fibre is again carried out. The rotation up to within say 30° of the unstable point is carried out uniformly and not very slowly. Then the crystal is allowed to settle down and when it is no longer oscillating the torsion head is very slowly turned in a series of small steps. Near the unstable position it will be necessary to wait a short time to give the crystal a chance to start spinning if it has already passed the unstable position. The reading of the torsion head at this position is ϕ.

2.2.5. *Example*

Method A

A rectangular parallelepiped of calcite was used and from its dimensions and weight and the known specific gravity, 2·72, the relation $I/v = 0{\cdot}128$ was obtained. The observations on the time of oscillation in different strengths of magnetic field are given in Table 2.2.

TABLE 2.2

Relation between magnetic field strength and time of oscillation

Magnetic field H (œ)	T_1 (s)	$\frac{1}{T_1^2} - \frac{1}{T_0^2}$	$\chi_1 - \chi_2$ (c.g.s.e.m.u./cm³)	$\chi_1 - \chi_2$ (c.g.s.e.m.u./g)
0	12·8			
2165	2·96	0·108	$11{\cdot}6 \times 10^{-8}$	$4{\cdot}28 \times 10^{-8}$
2465	2·69	0·132	11·0	4·04
2640	2·54	0·149	10·8	3·97
		Mean	11·1	4·10

(Value given by Krishnan, Guha, and Banerjee (1933): $11{\cdot}1 \times 10^{-8}$ c.g.s.e.m.u./cm³ or $4{\cdot}08 \times 10^{-8}$ c.g.s.e.m.u./g.)

Method B

The following observations were made using an electromagnet. The field strength was measured with a fluxmeter. The results are given in the following table.

TABLE 2.3

Relation between the magnetic field strength and the twist of the torsion head, at the unstable position

Magnetic field H (œ)	Twist of torsion head $\theta - 45°$	$\chi_1 - \chi_2$ (c.g.s.e.m.u./cm³)	$\chi_1 - \chi_2$ (c.g.s.e.m.u./g)
0	0		
2165	465°	$10{\cdot}7 \times 10^{-8}$	$3{\cdot}93 \times 10^{-8}$
2465	575°	10·2	3·75
2640	655°	10·1	3·71
	Mean	10·3	3·80

(Value given by Krishnan, Guha, and Banerjee (1933): $11{\cdot}1 \times 10$ c.g.s.e.m.u./cm³ or $4{\cdot}08 \times 10^{-8}$ c.g.s.e.m.u./g.)

2.3. The determination of absolute magnetic susceptibilities in diamagnetic and paramagnetic crystals

When placed in a magnetic field of strength H, dia- and para-magnetic crystals develop an induced magnetic moment M, which is proportional to H. We may write

$$M = vH\chi$$

where v = the volume of the crystal, χ = susceptibility per unit volume, and H = field strength.

The value of χ varies with the direction in which the field is applied and the problem is to determine the principal susceptibilities χ_1, χ_2, χ_3 and their orientation, when that is not fixed by the symmetry. The simplest method of determining an absolute value of χ is to place the crystal in a non-uniform magnetic field with a principal axis along the lines of force and measure the tractive force exerted upon the crystal by the field. A convenient arrangement of the specimen and the magnet is indicated in Fig. 2.2. The crystal X is free to move in a horizontal direction ab in the plane bisecting the angle between the faces of the two pole pieces. The magnitude of the force F acting on the specimen is given by the relation

$$F = v\chi H\frac{\partial H}{\partial x},$$

where H = field strength at the centre of the specimen, and $\partial H/\partial x$ = rate of variation of H in the direction of movement of the specimen.

If the value of H at points along the line *ab* is measured and a plot of H against x is made, from the graph $\partial H/\partial x$ can be found at a selection of points. The graph representing $H\,\partial H/\partial x$ can then be constructed. This graph has a maximum; $\partial H/\partial x$ is zero somewhere near the strongest part of the magnetic field and H is zero at the considerable distance from the pole pieces. It follows that somewhere between $H\,\partial H/\partial x$ must reach a maximum value. As the repulsive force exerted on the crystal is a maximum at this point the measurements are carried out so as to utilize this fact.

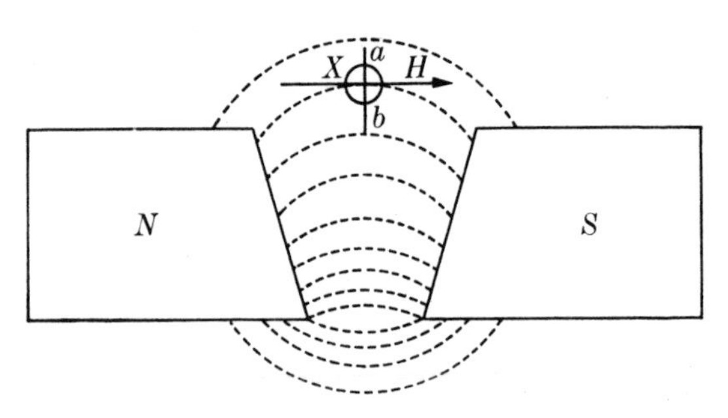

FIG. 2.2. Diagram showing the simplest relation between the direction of the magnetic field, *H*, and the direction of the tractive or repulsive force *ab*, when a crystal, *X*, is placed in an inhomogeneous field.

2.3.1. *Experimental arrangement*

The force exerted on the crystal by the magnetic field is under normal conditions not more than a few milligrams weight. It is therefore necessary to provide a sensitive device that can also constrain the crystal to move along the plane of symmetry of the magnetic field. This is achieved by the apparatus shown diagrammatically in Fig. 2.3. The crystal, *C*, is mounted on a copper wire, *W*, which is supported in a thin tube, *T*, attached to the plastic frame *F*. This frame has an axle mounted in ball races *BB* which in turn are supported by a pillar *P* attached to the body of a microscope *M*. The microscope can be racked towards or away from the pole pieces *NS* of an electromagnet. The wheel, *H*, driving the rack is rigidly fixed to the yoke of the electromagnet. (For the sake of clarity the coils of the electromagnet are omitted from Fig. 2.3(a), (b), (c).) Thus by turning the wheel *H* the whole assembly comprising microscope and frame carrying the crystal can be moved in or out of the regions where the magnetic field is strong.

The position of the crystal relative to the field is determined by an index *I*, which is attached to the frame *F*. This rotates with the frame, *F*, about the axis *BB*, i.e. it moves in the direction of the axis of the microscope. In order to measure the displacement of the index *I* its image in a mirror,

FIG. 2.3(d). Photograph giving a general view.

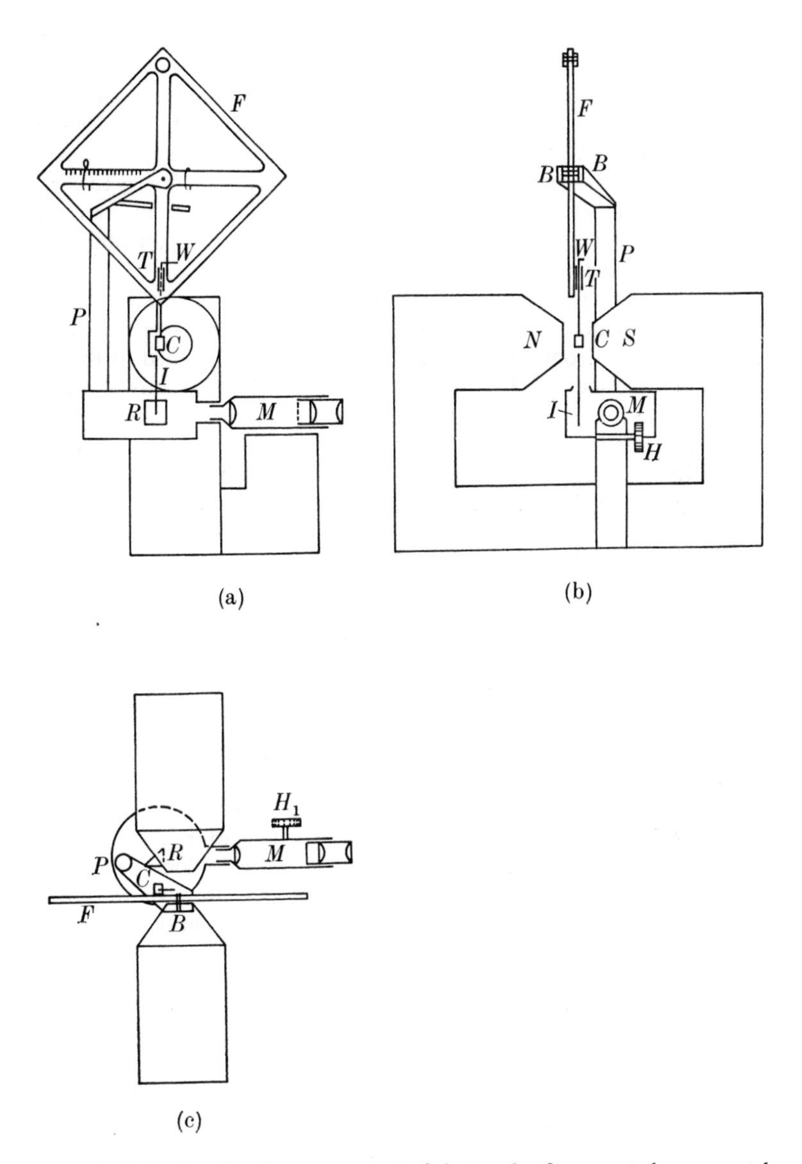

FIG. 2.3. Apparatus for the measurement of the tractive force exerted on a crystal in an inhomogeneous magnetic field (a) side view, (b) front view, (c) plan, (d) photograph giving a general view.

R, set at 45° to the direction of movement, is viewed by the microscope. The image of the index in the mirror moves in a plane perpendicular to the axis of the microscope and when once brought into focus it remains so. The mirror is a right-angled glass prism. The gap between the pole-pieces is made as small as convenient, so as to make the maximum value of $H\,\partial H/\partial x$ as great as possible. Care must be taken to ensure that the variation of $H\,\partial H/\partial x$ over the volume of the crystal is not serious. It is easy to check how great is this variation by measuring the couple exerted by the magnetic field with the microscope racked to positions close to, but on either side of, that giving the maximum force.

Two riders and a balance weight are used on the frame F. One rider is on the graduated arm on one side of the axle and the other rider is on the opposite horizontal arm. The balance weight is attached to the top corner of the frame. This balance weight serves to bring the centre of gravity of the whole rotating part near to the axis of rotation. The sensitivity of the system can be increased by adjusting the balance weight, but, if the sensitivity is increased too much, the time of oscillation becomes inconveniently long. A time of swing of a few seconds is suitable.

2.3.2. *Conduct of the experiment*

The crystal is mounted on the frame, F, with the trigonal axis (which in bismuth is perpendicular to the cleavage face) along the lines of force. The crystal must also lie in the plane of symmetry of the magnetic field that runs perpendicular to the lines of force.

The rider on the right-hand arm of the frame is set to zero on the scale. The rider on the left-hand arm is adjusted to bring the image of the index to the centre of the graticule in the eyepiece of the microscope M. The balance weight at the top of the frame F is adjusted so that the time of swing of the frame is a few seconds. The ball bearings must be free so that the frame performs several oscillations before coming to rest.

The magnetic field is now switched on and the deflection of the index noted as the microscope is racked towards or away from the region of strongest magnetic field. If the index goes out of the field of view of the microscope, the rider should be moved on the right-hand arm and adjusted to bring the image of the index back again. It will be found that the force exerted by the magnetic field reaches a maximum when the microscope is in a certain position. This is the position in which the measurements are made. The right-hand rider is moved to bring the image of the index to the centre of the field of view, first with the crystal set with the trigonal axis parallel to the lines of force, and then with the same axis perpendicular to the lines of force. If the displacements of the rider are d_3 and d_1 for these two settings then the ratio of the forces F_3/F_1 exerted by the magnetic field on the crystal in these two orientations is equal to d_3/d_1. Now

$$F_3 = v\chi_3\,H\,\partial H/\partial x,$$
$$F_1 = v\chi_1\,H\,\partial H/\partial x,$$

hence

$$F_3/F_1 = \chi_3/\chi_1 = d_3/d_1.$$

The bulk value of the susceptibility, χ_m, is available from published tables and is related to χ_1 and χ_3 by the expression,

$$\chi_m = \tfrac{1}{3}(\chi_3 + 2\chi_1).$$

From this expression and the ratio χ_3/χ_1 the individual values of χ_1 and χ_3 can be found.

If the value of $H\,\partial H/\partial x$ can be found with sufficient accuracy it is possible to find χ_1 and χ_3 directly without making use of the value of χ_m. It is, however, rather difficult to obtain Hall probes or search coils which are small enough and accurate enough to give satisfactory results.

2.3.3. *Example*

Strength of magnetic field measured in a fluxmeter = 8000 œ. After balancing the frame supporting the crystal, so that the image of the index was at zero when the field was not applied, the rider on the horizontal arm of the frame was moved to the following distances from the axle supporting the frame:

magnetic field perpendicular to trigonal axis, 7·9 cm;

magnetic field parallel to trigonal axis, 4·8 cm.

Hence

$$\frac{\chi_1}{\chi_3} = \frac{7{\cdot}9}{4{\cdot}8} = 1{\cdot}65.$$

From the published figure for χ_m, the susceptibility measured on a powder of bismuth, we have,

$$\chi_m = \tfrac{1}{3}(\chi_3 + 2\chi_1) = -13{\cdot}2 \times 10^{-6} \text{ c.g.s. e.m.u./cm}^3.$$

From the last equation and the ratio of χ_1/χ_3 we obtain

$$\chi_1 = -15{\cdot}2 \times 10^{-6}; \qquad \chi_3 = -9{\cdot}2 \times 10^{-6} \text{ c.g.s. e.m.u./cm}^3.$$

3. Thermal Properties of Crystals

3.1. Conduction and expansion

WHEN heat is applied at a point within a single crystal it is conducted away from the point of application in all directions and gives rise to expansion or contraction. These properties are second-order tensor properties. Thermal conduction expresses the relation between the vector of heat flow and the vector of temperature gradient. Thermal expansion expresses the relation between the amount of expansion in a given direction and the original length in that direction. Being second-order properties they are represented according to the symmetry in the same way as dia- or para-magnetism (see section 2.1).

A distinction must be made between two ideal sets of conditions under which conduction takes place. In one an infinite plate of finite thickness is used and in the other a rod from which heat cannot escape in directions perpendicular to the length is used. The former case determines what is known as thermal conductivity, the latter what is known as thermal resistance. In an anisotropic body, for which the general triaxial ellipsoid is the appropriate representation surface, the directions of maximum temperature gradient and of heat flow only coincide in special directions, for example along one of the principal axes of the ellipsoid. In general

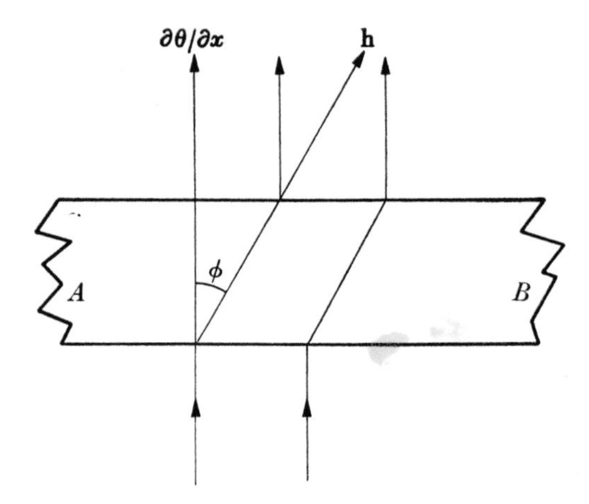

FIG. 3.1. Diagram for the flow of heat through a thin extended plate.

directions there is a greater or smaller angle between the two vectors **h** and $\partial\boldsymbol{\theta}/\partial\mathbf{x}$ representing heat flow and maximum temperature gradient respectively. Figure 3.1 shows the relation applying to an infinite plate *AB* and Fig. 3.2 the corresponding relation for a rod *CD*. In Fig. 3.1 the planes of constant temperature, the so-called isothermal surfaces, must all be parallel to the plane of the plate *AB*. The maximum temperature gradient must be perpendicular to these isothermal surfaces and hence the vector $\partial\boldsymbol{\theta}/\partial\mathbf{x}$ is perpendicular to the plane of the plate. In Fig. 3.2 the heat must necessarily travel in straight lines parallel to the length of the rod. Both of these arrangements only determine one of the two quantities involved

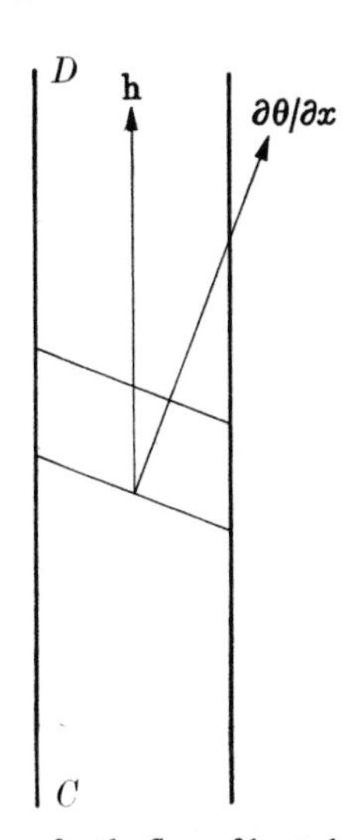

FIG. 3.2. Diagram for the flow of heat along a narrow rod.

and the other is left to any orientation which the degree of anisotropy imposes on it.

3.2. Coefficients of thermal conductivity and thermal resistance

The general relation between the two vector quantities **h** and $\partial\boldsymbol{\theta}/\partial\mathbf{x}$, representing heat flow and temperature gradient in a direction **x** is the same as applies to all second-order tensor properties. We may write

$$h_i = k_{ij}(\partial\theta/\partial x_j) \tag{3.1}$$

or

$$(\partial\theta/\partial x_i) = r_{ij}\,h_j. \tag{3.2}$$

The coefficients k_{ij} are called thermal conductivities and the coefficients r_{ij} are known as thermal resistances. In using an infinite plate of finite thickness the direction and magnitude of the maximum temperature gradient are defined, i.e. we know $\partial\theta/\partial x_1$, $\partial\theta/\partial x_2$ and $\partial\theta/\partial x_3$. Thus given the k_{ij}'s we can calculate the h_i's but we can only measure $\sqrt{(h_1^2+h_2^2+h_3^2)}$. If a rod of finite area is used we can measure the total heat flow and, knowing the crystallographic orientation of the axis of the rod we can obtain h_1, h_2 and h_3. Then if we know the r_{ij}'s we can calculate the $(\partial\theta/\partial x_i)$'s. From the components of the latter vector we can calculate the orientation of the isothermal surfaces with respect to the axis of the rod. If the axes of reference are chosen parallel to the principal axes of the triaxial ellipsoid then

$$k_{ii} = \frac{1}{r_{ii}} \qquad i = 1, 2, \text{ or } 3$$

and

$$k_{ij} = 0 \qquad r_{ij} = 0 \quad i \neq j.$$

3.3. Measurement of thermal conductivity using flat plates

The thermal conductivities of bad conductors of heat are difficult to determine with reasonable accuracy, but one of the best methods is a modification of Forbes's divided bar method. In this method heat flow is measured through a composite bar consisting of metal, bad conductor, metal, the cross-section of all being the same. The dimensions are so chosen that practically all the heat flows through the composite bar from one end to the other and little is lost from the sides. Under this condition the heat flow H is the same in the metal and in the bad conductor and we may write

$$H = k_{\mathrm{m}} A(\partial t/\partial x)_{\mathrm{m}} = k_{\mathrm{c}} A(\partial t/\partial x)_{\mathrm{c}}$$

or

$$k_{\mathrm{c}}/k_{\mathrm{m}} = (\partial t/\partial x)_{\mathrm{m}}/(\partial t/\partial x)_{\mathrm{c}},$$

where k_{c} and k_{m} are respectively the conductivities of the crystal and the metal and $(\partial t/\partial x)_{\mathrm{c}}$ and $(\partial t/\partial x)_{\mathrm{m}}$ are the corresponding temperature gradients. A is the area of cross-section. The conductivity of the metal is known and hence that in the crystal can be found.

There are certain conditions that are difficult to satisfy. For the above theory to apply there must be a negligible loss of heat from the sides of the crystal and metal blocks. In the present experiment this is largely achieved by enclosing the divided bar in an extremely poor conductor, namely, expanded polystyrene. Another requirement is that the heat flow through the blocks shall be uniform over the whole area of cross-section. This depends on a number of factors. The heat is put in by a heated tube of small diameter at one end of the hot block. The length of the block between the heating element and the crystal is made long enough to ensure that the temperature across the block where it is cemented to the crystal is uniform. Similarly at the cooled end the heat is removed by water flowing through two tubes which pass through the block at one end. The lines of heat flow should be parallel to the length of the block and by using two tubes at right angles to the length of the block and a sufficient length of block this condition is sufficiently fulfilled. A much more difficult matter is presented by the insulation of the cement used to attach the metal blocks to the crystal. The material used in this apparatus is an epoxy resin known as Araldite. Its thermal conductivity is probably ten times less than that of the crystal and hence the uniformity of the heat flow across the block is closely dependent on the uniformity of the thickness of these layers of cement. When the resin is applied the blocks are clamped together so that the layer should be uniform, but a small departure from uniformity is difficult to discover and it would seriously affect the results.

3.3.1. *Experimental details*

Two similar boxes are provided, each containing heated and cooled metal blocks between which blocks of quartz are cemented. One block is cut with its shortest edge—the direction of heat flow—parallel to the trigonal axis of the crystal. The other block is cut with this edge perpendicular to the trigonal axis. In all other respects the two divided bars are the same. Holes have been drilled in the metal and crystal blocks and chromel–alumel differential thermocouples inserted in them. The thermocouples in the quartz have been cemented in place using Araldite and the couples in the metal blocks have been insulated from the metal. *These couples should not be removed from the holes.*

The measurement of the temperature differences between two holes in the quartz and two holes in the cooled metal block is effected by the Comark electronic thermometer. The four pairs of thermocouple leads—two from each divided bar—are connected to the terminals marked 1, 2, 3, 4 at the back of the Comark thermometer. The left-hand knob on the front of the instrument selects the differential thermocouple and the right-hand knob selects the temperature range. Before every measurement the zero-set (black) button should be depressed—when the needle will return nearly to zero—and the white knob above it should be rotated until the needle is exactly at zero. This is a very important adjustment which is not the same for the different ranges of temperature. The instrument is calibrated for the chromel–alumel thermocouples used in the divided bars.

To make satisfactory measurements it is necessary that the rates of heating and cooling should remain constant for sufficiently long to establish equilibrium conditions. The uniformity of heating is not usually a problem since it depends only on the constancy of the mains potential. The cooling presents more of a problem owing to the frequent fluctuations in water pressure. To overcome this a constant-flow valve is provided for inclusion between the mains tap and the apparatus. This valve has an adjustable knob that may be used to regulate the flow. There is no particular rate of flow that is necessary. The important condition is that is should be constant. A rate of flow of a litre in 1–2 min is suitable.

Equilibrium conditions cannot be established quickly because it is necessary for the insulation surrounding the divided bar to attain its final temperature. A control over this approach to equilibrium can be made by measuring the ratio of the temperature differences in the crystal and metal of either divided bar as a function of time.

3.3.2. *Procedure*

Before starting the heating and cooling it is convenient to measure, with a pair of callipers, the distance between the holes in the quartz blocks and in the cooled metal blocks. (The heated metal blocks have a less uniform flow of heat through them than the cold blocks and are therefore not used to determine the temperature gradient). Care must be taken not to disturb the thermocouples in making this measurement.

The boxes are then closed and they are not again opened during the measurements—the water flows through the two boxes in series. The heating is intended for a 115-V supply. The heating elements have a resistance of about 2000 Ω. The thermocouple leads are connected to the terminals on the electronic thermometer and measurements can be made until equilibrium has been established. The conductivity of the metal is 0·20 cal/s cm °C. If the temperature differences in the quartz and metal are Δt_q and Δt_m and the distances apart of the thermocouples are x_q and x_m respectively, then the conductivity of the quartz is given by k_q where

$$k_q = \frac{\Delta t_m}{\Delta t_q} \cdot \frac{x_q}{x_m} 0{\cdot}20 \text{ cal/s cm °C.}$$

EXAMPLE

Two composite bars contained (a) a quartz block 25 × 25 × 15 mm cut with the optic (trigonal) axis perpendicular to the major faces and (b) a block of the same dimensions cut with the trigonal axis parallel to the major faces.

The distance apart of the holes in the quartz = 1·05 cm.

The distance apart of the holes in the brass = 1·5 cm.

(a) The temperature difference between the holes in the quartz = 10·8°C.

The corresponding difference between holes in the brass = 1·75°C.

Taking the conductivity of the brass as 0·20 we obtain

$$k_3 = 0{\cdot}2 \times \frac{1{\cdot}75}{10{\cdot}8} \times \frac{1{\cdot}05}{1{\cdot}5} = 0{\cdot}023 \text{ cal. cm}^{-1}\text{ s}^{-1}\text{ °C}^{-1}.$$

(b) The temperature differences between the holes:

in quartz is 12·0°C,

in brass is 1·13°C.

Thus

$$k_1 = 0{\cdot}2 \times \frac{1{\cdot}13}{12{\cdot}0} \times \frac{1{\cdot}05}{1{\cdot}5} = 0{\cdot}013 \text{ cal. cm}^{-1}\text{ s}^{-1}\,{}^\circ\text{C}^{-1}.$$

(Kaye and Laby, *Physical and chemical constants*

$$k_1 = 0{\cdot}0129; \qquad k_3 = 0{\cdot}0222.)$$

3.4. The measurement of the ratio of the principal thermal conductivities by the twin-plate method

3.4.1. *Principle of the method*

The thermal conductivity of a crystal varies with the direction of heat flow relative to the crystallographic axes. In general, the direction of heat flow through a small element of volume of the crystal is inclined to the direction of the maximum temperature gradient in that element. Two different types of measurement are therefore possible, depending on the shape of the piece of crystal studied. If the crystal is long in comparison with its breadth and thickness, and the heat flows along its length, the direction of heat flow must be parallel to its length. The isothermal surfaces, i.e. the surfaces over which the temperature has a given value, are in this case not necessarily perpendicular to the length of the crystal. At the ends, where metal surfaces apply the heating and cooling, the isothermal surfaces must be parallel to the heated or cooled plates, but away from the ends of the crystal plate the isothermal surfaces take up the appropriate orientation corresponding to the direction of heat flow.

When a thin plate of crystal is held between two metal surfaces, one heated and the other cooled, the isothermal surfaces in the crystal are, except near the outer edges, parallel to its large faces and the maximum temperature gradient is perpendicular to the heated and cooled surfaces. But in this case the direction of heat flow is not determined and, in general, is not normal to the large faces.

The direction of heat flow and the direction of the maximum temperature gradient are the same for isotropic (cubic) crystals and for rods and sections parallel or perpendicular to the principal crystallographic axis of uniaxial crystals. In this experiment use is made of bars of quartz elongated in a direction inclined at about 45° to the trigonal axis, which lies in the plane of the crystal plate.

It may be shown that the equation to an isothermal surface surrounding a point source of heat may be written

$$\frac{x_1^2}{k_1} + \frac{x_2^2}{k_2} + \frac{x_3^2}{k_3} = \text{constant},$$

where x_1, x_2, x_3 are the coordinates of a point on the isothermal surface and k_1, k_2, k_3 are the principal thermal conductivities. For our experiment we may put $k_1 = k_2$ since the crystal is trigonal. The plate is cut parallel to a plane containing the X_3 axis and so we may put $x_2 = 0$. Thus the equation reduces to

$$\frac{x_1^2}{k_1} + \frac{x_3^2}{k_3} = \text{constant}.$$

This is the equation to an ellipse having semi-axes proportional to $\sqrt{k_1}$, $\sqrt{k_3}$, respectively. In Fig. 3.3 the heat is supposed to flow in the direction OP through the point P having coordinates (x_1, x_3) and the axis $OA = \sqrt{k_3}$ is taken parallel to the trigonal axis. The axis $OB = \sqrt{k_1}$ may be any direction perpendicular to OA. It may be shown that the angle ϕ between

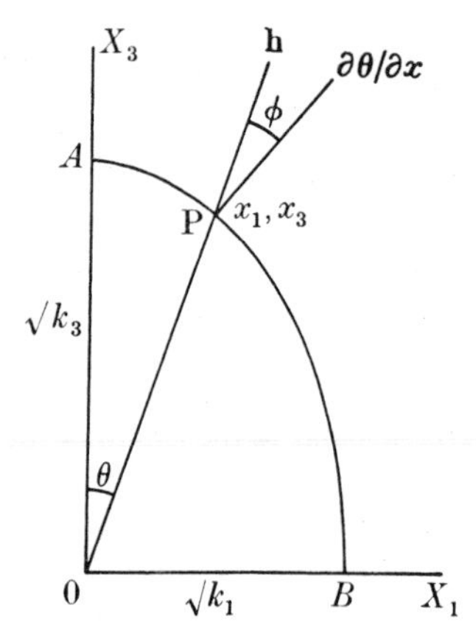

FIG. 3.3. Diagram showing the relation between the angle ϕ, between the vectors of heat flow and of the maximum temperature gradient, and the isothermal surface AB.

the normal at P to the isothermal surface and the direction of heat flow, OP, is given by

$$\tan(\theta + \phi) = \frac{k_3}{k_1} \tan\theta.$$

Thus knowing θ and measuring ϕ, the value of k_3/k_1 may be determined.

3.4.2. *Experimental arrangement*

To avoid disturbance of the isothermal surface by cooling from the side of the plate, two plates are placed side by side. They are cut from the same piece of quartz so that the inclination of the length of the plate to the principal (trigonal) axis is the same. In mounting the plates in the apparatus one is turned over relative to the other, so that along the line where the two plates adjoin the inclination of the isothermal surfaces to one another is 2ϕ. This use of two crystal plates with a twin-like mutual orientation has given rise to the name 'twin-plate experiment', Fig. 3.4.

Heat is produced in a fixed block into which one end of each plate is pressed and another water-cooled block supports the other ends of the plates. The isothermal surfaces are made visible by the use of elaidic acid. This white crystalline powder melts at 44°C and is sprinkled on the surface, or, if already molten, is lightly brushed over the surface. The boundary between the molten and unmolten elaidic acid can be seen with appropriate illumination. A transparent plastic plate having a number of parallel lines ruled on it is pivoted above the crystal plates. The setting of this set of lines can be read on a scale engraved on the transparent sheet. When the isothermal surface corresponding to the melting point of elaidic acid can be seen the lines are set parallel, first to the boundary on one plate and then parallel to that on the other plate. The angle between these two settings is read off on the scale.

3.4.3. *Conduct of the experiment*

It is important to remember that quartz plates are easily chipped and if dropped are very liable to shatter. The experiment should be performed twice; once with the plates arranged so that the faces in contact are hotter than the outer parallel faces at the same distance from the heated block and again with the plates changed over so that the inner faces are cooler than the outer faces. Some practice is necessary in securing a good boundary between the molten and unmolten elaidic acid. One method that has given good results is to have a small tray in which a little acid is kept molten by placing it on the heated block and to use a small brush to paint this molten acid in a thin layer over the crystal plates. It may be found helpful to cut two pieces of paper and lay them on the plates parallel to the isothermal surfaces. It is then easier to bring the lines engraved on the perspex parallel to the isothermal surfaces.

EXAMPLE

Values of the angle 2ϕ, measured on successive deposits of elaidic acid were 34°, 35°, 34°, 31°, 34°.

With crystals reversed in the apparatus the values obtained were 34°, 33°, 34°. Mean 33°. $\theta = 45°$. $\tan(\theta+\phi) = \tan 61{\cdot}5 = 1{\cdot}8(4) = k_3/k_1$. (Kaye and Laby, *Physical and chemical constants* $k_3/k_1 = 1{\cdot}72$.)

3.5. The study of isothermal surfaces on crystal plates

3.5.1. *Introduction*

We imagine a point source of heat to exist inside a crystal and we consider a closed surface surrounding the point source which is at a constant temperature. This is called an isothermal surface. In cubic crystals the isothermal surfaces are concentric spheres. In tetragonal, trigonal and hexagonal crystals they are ellipsoids of revolution, the axis of revolution being the principal crystallographic axis. In orthorhombic, monoclinic, and triclinic crystals the isothermal surfaces are triaxial ellipsoids, the principal axes of which coincide with diad axes in the crystals when these are present.

A small hole is drilled in a thin parallel-sided plate of crystal so that a heated rod may be inserted into the plate. The change of temperature at points on the plate may be observed by coating the plate with elaidic acid and noting the boundary between the molten and the unmolten region. Ideally the crystal should be infinitely large to avoid effects due to the edges. Ideally also the isothermal surface should correspond to a steady state, i.e. it should have remained stationary for some time before the heating of the hot point was turned off. A heated point is impracticable and a compromise has to be made by using a rod of diameter small compared with the size of the crystal.

Any central section of an isothermal surface is an ellipse the major and minor axes of which are proportional to the square roots of the corresponding principal conductivities.

3.5.2. *Apparatus*

The apparatus consists of a heated rod mounted horizontally. A metal disk mounted on the same base, but separate from the heated rod, serves as a radiation shield. Its function is to prevent heat reaching the crystal plate except through the heated rod (Fig. 3.5).

The crystal plates are of quartz, SiO_2, cut parallel to (0001), $(10\bar{1}1)$, $(10\bar{1}0)$ and a plate of gypsum, $CaSO_4 . 2H_2O$, cleaved parallel to (010).

3.5.3. *Procedure*

Each plate should be heated enough so that a uniform molten layer of elaidic acid can be formed all over it. The plate is then allowed to cool down. When it is again at room temperature it is threaded on to the heated rod, care being taken not to damage the plate in doing so. The spreading of the isothermal surface should be noted. It should not be too rapid as otherwise the static isothermal surface does not correspond with the observed boundary. When the melting has gone far enough to permit the boundary to be conveniently studied, the crystal is removed from the heated rod and allowed to cool down.

The major and minor axes of the ellipses can be measured either by callipers or on a travelling microscope. The quartz plate cut parallel to $(10\bar{1}1)$ does not give elliptical figures because the principal axes do not lie in the plane of the plate. The directions, relative to the $(10\bar{1})$ cleavage cracks, of the major and minor axes of the ellipse obtained with the cleavage plate of gypsum should be noted.

3.6. Measurement of the principal values of thermo-electric e.m.f.s in crystals

3.6.1. *Introduction*

When two dissimilar metal wires are joined at two points and one junction is heated relative to the other a thermo-electric electro-motive force is generated. A current circulates round the two wires so long as the difference in temperature is maintained. The same phenomenon occurs when a block of conducting single crystal is held in contact with two pieces of

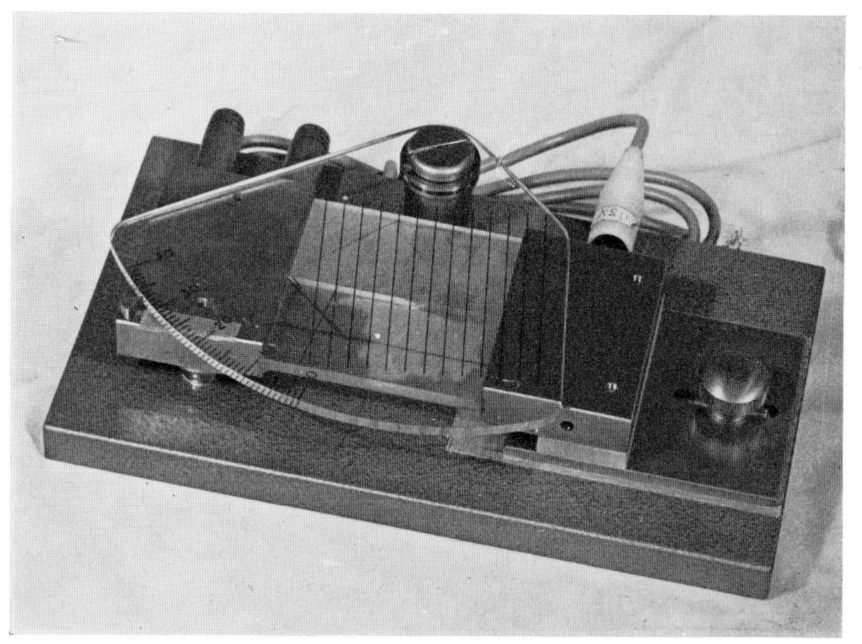

FIG. 3.4. Apparatus for 'twin-plate' experiments, to determine the ratio of the principal thermal conductivities in quartz.

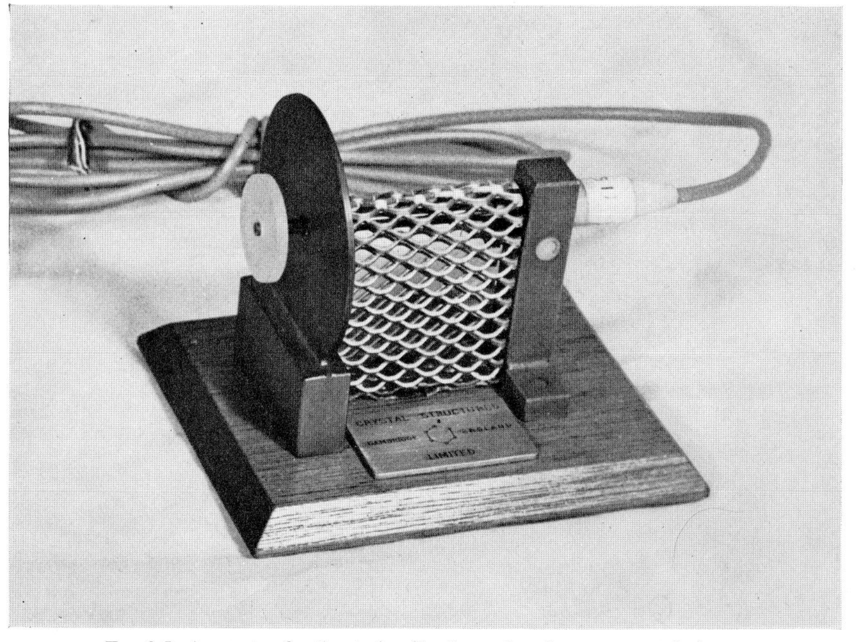

FIG. 3.5. Apparatus for the study of isothermal surfaces on crystal plates.

the same metal which are at different temperatures and are joined by a wire of the same material. The magnitude of the e.m.f. generated in this way depends on the orientation of the crystal relative to the direction of heat flow. We are here concerned with the relation between two vector quantities, namely, the temperature gradient and the flow of electric current. As in all cases where two vector quantities are related in a crystal the coefficients relating these quantities form a second-order tensor. Thus the surface that may be constructed by drawing lines from a given point of length proportional to the ratio of the electric current (or potential) to the temperature gradient in the same direction, is an ellipsoid. Further, it is a triaxial ellipsoid for triclinic, monoclinic, and orthorhombic crystals, an ellipsoid of revolution for hexagonal, trigonal, and tetragonal crystals, and a sphere for cubic crystals.

Just as is the case with thermal conductivity the directions of maximum temperature gradient and of electric current flow do not in general coincide. Only along directions parallel to the principal axes of the ellipsoid is this the case. Thus in hexagonal and trigonal crystals the direction of the trigonal axis is one such direction and all directions perpendicular to this axis are also characterized in the same way. Thus for pieces of crystal cut in directions inclined to the principal axis there are transverse thermo-electric e.m.f.s, with which, however, we shall not be concerned here.

3.6.2. *Experimental arrangement*

Crystals are cut so that they may be clamped between two brass blocks, one of which is heated and the other cooled. The blocks are joined to a galvanometer for reading the potential difference between the two blocks. Holes in the blocks permit the insertion of a thermometer for reading the temperatures. If the galvanometer gives a deflection corresponding to S μA and the resistance in the circuit is $R\Omega$ then the thermo e.m.f. is RS μV. If t is the temperature difference between the two metal blocks, the thermo-electric power of the crystal under test is $RS \times 10^{-6}/t$ V/°C.

3.6.3. *Procedure*

Crystals of bismuth and tellurium are provided. The bismuth block has a cleavage face which is perpendicular to the trigonal axis. This face can be recognized by its smoothness. The crystal is clamped between the blocks first with the cleavage faces in contact with the blocks. The measurement carried out as described above gives the thermo-electric power in a direction parallel to the trigonal axis. The crystal is now turned so that the trigonal axis is parallel to the surfaces of the blocks and the measurement is repeated.

The crystal of tellurium has three cleavages at 60° to one another and these planes are all parallel to the trigonal axis. Mount the crystal first with the trigonal axis perpendicular to the faces of the blocks and then with this axis parallel to these faces.

Difficulties arise with many semiconducting crystals on account of the poor electrical contact made with the metal blocks. If the resistance of the specimen is appreciable in comparison with the (shunted) resistance of the galvanometer it must be determined. This can be done by including a resistance box in the circuit and finding the change in the galvanometer deflexion due to the insertion of a known resistance in the circuit. If R_g, r_c, and r are the resistances of the galvanometer, crystal and resistance box respectively, e the thermo-e.m.f. generated between the metal blocks, g the current through the galvanometer, then

$$g = e/(R_g + r_c + r).$$

By making several determinations of corresponding values of g and r the values of e and r_c can be found.

3.7. Thermal expansion of crystals

3.7.1. *Introduction*

When crystals are heated they may expand in all directions or contract in all directions or expand in some directions and contract in others. We are here concerned with two vectors which are (a) the change in length for a change in temperature of 1°C of a line in the crystal and (b) the vector representing that line in length and direction. These two vectors will be denoted **p** and **r** respectively. In general there is a relation between these vectors which can be expressed

$$p_i = \alpha_{ik} r_k.$$

This equation implies that each component of the vector **p** is proportional to each component of the vector **r**. The coefficients of expansion α_{ik} form a second-order tensor having the same transformation properties as the conductivities. The representation surface is an ellipsoid or an hyperboloid. It is quite possible with the αs to obtain negative values and in this case there are certain directions of zero expansion.

3.7.2. *Measurement of thermal expansion coefficients using X-rays*

Direct measurement of the expansion of crystals is difficult to make since the amount of expansion is so small. X-ray methods of measuring expansion, however, are applicable to very small crystals, and are quite simple to use.

Since the spacing d of a set of planes in a crystal is related to the glancing angle θ and the wavelength λ by the equation

$$d = \lambda/(2\sin\theta)$$

the expansion of d on raising the temperature may be measured by noting the change in θ.

Differentiating d with respect to θ we have

$$\partial d/\partial\theta = -\tfrac{1}{2}\lambda.\operatorname{cosec}\theta.\cot\theta = -d\cot\theta$$

or

$$\frac{\partial d}{d} = -\cot\theta.\partial\theta.$$

The thermal expansion coefficient α is defined by the equation

$$\alpha = \frac{\partial d}{d} \cdot \frac{1}{\partial t},$$

where ∂t is the change of temperature producing the change of spacing ∂d. Thus we have

$$\alpha = -\cot\theta \,.\, \partial\theta/\partial t.$$

From this equation it will be seen that for a given change of spacing ∂d, $\partial\theta$ increases from a finite to an infinite value as θ approaches 90°. Thus by reflecting the monochromatic X-rays almost normally to the crystal face the highest separation of the reflections at two temperatures is obtained. The apparatus used to achieved this result is called a back-reflection camera, and is shown in Fig. 3.6. The slit S for the incident beam passes through the photographic film F enclosed in the camera C. This receives the X-rays

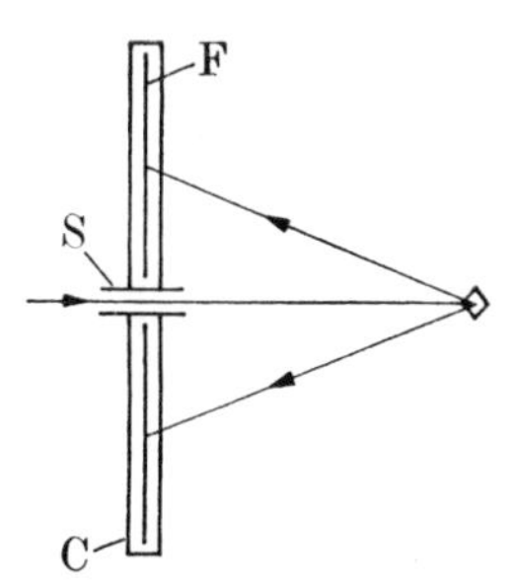

FIG. 3.6. Diagram of back reflection X-ray camera.

that are reflected back almost along their incident path. The crystal is oscillated so that the reflected beam falls first on one and then on the other side of the slit.

If the distance between the pairs of corresponding spots is $2x$, and r is the perpendicular distance from crystal to film, then

$$\frac{x}{r} = \tan(\pi - 2\theta),$$

where 2θ is the deviation of the X-rays. Thus θ may be found by measuring x and r.

If d_1 and d_2 are the spacings of a given plane at temperatures t_1 and t_2,

$$d_1 = \frac{\lambda}{2\sin\theta_1} \qquad \text{and} \qquad d_2 = \frac{\lambda}{2\sin\theta_2}.$$

The definition of the coefficient of expansion α is

$$d_2 = d_1\{1 + \alpha(t_2 - t_1)\} \qquad \text{or} \qquad \alpha = \frac{d_2 - d_1}{d_1(t_2 - t_1)}.$$

Substituting d_1 and d_2,

$$\alpha = \frac{\frac{\lambda}{2}\left(\frac{1}{\sin\theta_2} - \frac{1}{\sin\theta_1}\right)}{(\lambda/2\sin\theta_1)(t_2 - t_1)} = \frac{\sin\theta_1 - \sin\theta_2}{\sin\theta_2(t_2 - t_1)}.$$

This is an exact expression that should be used when the change in θ, consequent on the expansion, is comparable with $(\pi - 2\theta)$. The formula for α obtained by differentiation, namely,

$$\alpha = -\cot\theta\,.\,\partial\theta/\partial t$$

is only valid provided that the change in θ is small in comparison with $(\pi - 2\theta)$. This condition is usually fulfilled.

3.7.3. *Experimental details*

If a flat back-reflection camera is used it may be mounted on an X-ray oscillation goniometer but the distance between the axis of oscillation and the plane of the film must be great enough to permit the heating apparatus to be placed round the crystal. The simplest device for heating the crystal consists of a tube directed towards the crystal leading a stream of air heated by passage through a metal coil kept at a constant temperature. The rate of flow of the air must also be kept constant. This apparatus hardly affects the normal functioning of the X-ray goniometer but there may be difficulty in determining the temperature of the crystal. If a small thermocouple can be fixed to, or near, the crystal a reliable measure of the temperature can be obtained. Alternatively, the crystal may be mounted in a block that surrounds it and can be raised to a known temperature.

It is not necessary to use a flat-plate camera; a Debye–Scherrer camera, of diameter not less than 19 cm, will be found suitable provided that means can be provided for heating the crystal. Special high-temperature powder cameras enabling the specimen to be heated to 1000°C or even 1500°C are manufactured but these are usually too elaborate for student use. Since the crystal may either expand or contract on heating it is necessary to take two photographs—one at room temperature (showing two spots, one on either side of the hole in the film), the other at the higher temperature. It may lead to ambiguity if the photographs at both temperatures are taken on one film. However, by making different exposure times it may be possible to distinguish between the pairs of reflections. Great care must be exercised in the selection of the plane of reflection and of the wavelength of the monochromatic radiation. Experimentally it is difficult to make θ greater than 85° because in that case the X-rays are reflected almost backwards along their own track. When θ is near to 90° there is an almost inverse relationship between $\partial\theta$ and $[(\pi/2) - \theta]$. This results in the value of $\partial\theta$ being rather small for practical use if θ is less than about 75°. There is thus quite a narrow useful range of θ, and consequently of d, for any given wavelength. This is one of the practical difficulties of the method. Every crystal has only a finite number of values of d and for the usual monochromatic radiations provided by the $K\alpha$, $K\beta$ lines of Cr, Co, Fe, Ni, Cu, Mo, and Ag it may be difficult to find a reflection of the most suitable kind which has θ lying in the range 75°–85°. The following table gives the range of values of the spacing, d, which are suitable for this measurement.

Radiation	λ (Å)	Range of d (Å) for θ between 75° and 85°
$FeK\alpha$	1·937	0·973–1·003
$CuK\alpha$	1·542	0·774–0·798
$MoK\alpha$	0·711	0·357–0·368

It is sometimes convenient to compare the separation of the α_1, α_2 doublet with the displacement of the spot on the film caused by the change in temperature. Differentiating the equation

$$\lambda = 2d \sin\theta$$

with respect to change in λ we obtain

$$\partial\lambda = 2d\cos\theta \, . \, \partial\theta$$

or

$$\frac{\partial\lambda}{\lambda} = \cot\theta \, . \, \partial\theta.$$

The value of $\partial\lambda/\lambda$ is well known and for Cu $K\alpha$ it is 0·00249. If the shift of the $K\alpha_1$ component towards a smaller θ-value, due to the thermal expansion is m, and the separation of the spots due to the α_1, α_2 components is n, then

$$\alpha/0{\cdot}00249 = m/n \, . \, \partial t$$

or

$$\alpha = 0{\cdot}00249 \, . \, m/n \, . \, \partial t.$$

It will be seen that this method does not depend on measuring the angle θ or the crystal-film distance.

EXAMPLES

(a) Debye–Scherrer photographs of silver at 18°C and 630°C.

Corresponding powder lines on a camera of radius 4·49 cm were distant from the point of emergence of the direct beam through the film by

125·6 mm at 18°C,

120·5 mm at 630°C.

Hence

$$\theta_{18^\circ} = 80^\circ\ 9',$$

$$\theta_{630^\circ} = 76^\circ\ 54'.$$

Using the formula

$$\alpha = \frac{\partial d}{d(t_2 - t_1)} = -\frac{\cot\theta\,.\,\partial\theta}{(t_2 - t_1)}$$

we obtain

$$\alpha = \tfrac{1}{2}(\cot 80^\circ\ 9' + \cot 76^\circ\ 54') \times 0{\cdot}0567 \times \tfrac{1}{612}$$

$$= 18{\cdot}8 \times 10^{-6}/^\circ\mathrm{C}.$$

Using the formula

$$\alpha = \frac{\sin\theta_{18^\circ} - \sin\theta_{630^\circ}}{\sin\theta_{630^\circ}(630 - 18)}$$

we obtain

$$= 18{\cdot}9 \times 10^{-6}/^\circ\mathrm{C}.$$

(Kaye and Laby, *Physical and chemical constants*: $18{\cdot}8 \times 10^{-6}/^\circ$C.)

(b) Back reflection photographs of single crystal of β-tin.

Reflection from 004. Spacing 0·7875 Å.

$$\text{Wavelengths CuK}\alpha_1 = 1{\cdot}54050\ \text{Å},$$

$$\text{CuK}\alpha_2 = 1{\cdot}54435.$$

$$\partial\lambda = 0{\cdot}00385,$$

$$\lambda_{\text{mean}} = 1{\cdot}54243,$$

$$\frac{\partial\lambda}{\lambda} = 0{\cdot}002496.$$

Distance apart of α_1, α_2 spots = 2·0 mm.
Displacement of spots due to temperature change = 1·4 mm.
Temperature change = 100°C.
Hence

$$\alpha_3 = \frac{1}{100} \times 0{\cdot}002496\,.\,\frac{1{\cdot}4}{2{\cdot}0} = 17.(5) \times 10^{-6}/^\circ\mathrm{C}.$$

4. Plastic Deformation

4.1. Fundamentals of plastic deformation

WHEN a crystal is subjected to a stress it is deformed, and if this deformation persists when the stress is removed it is said to be plastically deformed. Plastic deformation can take place in several ways and we shall here deal with two, namely, gliding and twinning. Of these, deformation by gliding is the commoner. Perhaps it is best exemplified in single crystals of certain metals. Deformation by twinning is well shown in calcite. When a crystal glides one part of it slips relative to a neighbouring part parallel to certain particular lattice planes, the slip being restricted to a certain direction. Plastic deformation by twinning differs from deformation by gliding in that parallel to the plane of slip there is the same movement of every lattice plane relative to its neighbours. Further, the extent of this slip cannot assume any arbitrary value but instead bears a definite relation to the cell size. The relative movement of any two planes is strictly proportional to their distance apart, hence the deformation may be said to be homogeneous. Deformation by gliding is usually inhomogeneous. The distinction between homogeneous and inhomogeneous deformation is shown in Fig. 4.1(a), (b). In Fig. 4.1(a) gliding has occurred in an inhomogeneous manner since most of the gliding has been confined to a few planes. Homogeneous gliding is illustrated by Fig. 4.1(b) where every plane has moved by the same amount relative to its neighbours. At ordinary temperatures plastic deformation is associated with cold-working, the name given to the process that leads to an increase in strength and hardness of the deformed crystal. On resting for some time without applied stress the crystal shows the phenomenon of recovery, i.e. loss of the increase of strength due to cold-working.

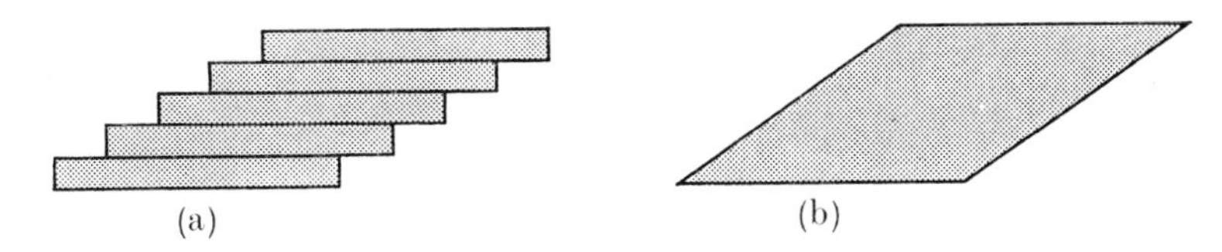

FIG. 4.1. Diagrams illustrating two types of deformation: (a) inhomogeneous, (b) homogeneous.

4.2. Homogeneous deformation

Let *ABCDE* on Fig. 4.2 represent a section of a sphere marked out in a crystal. If the crystal is subjected to a stress that gives rise to a homogeneous deformation by slipping parallel to the line *AD* on the plane which contains

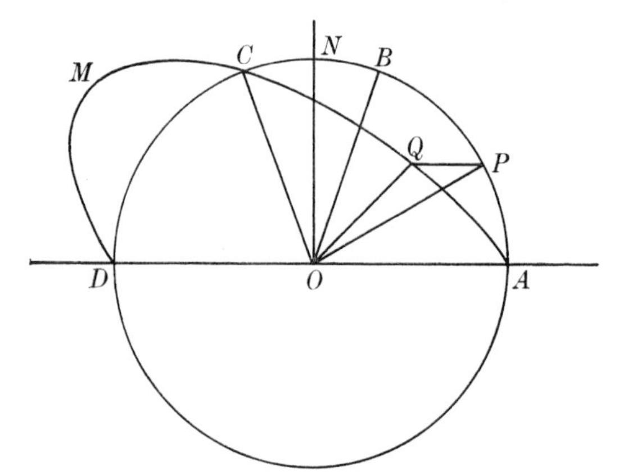

FIG. 4.2. Diagram showing the change of a circle, *AND*, into an ellipse, *ACMD*, as a result of homogeneous deformation (gliding parallel to *AD*).

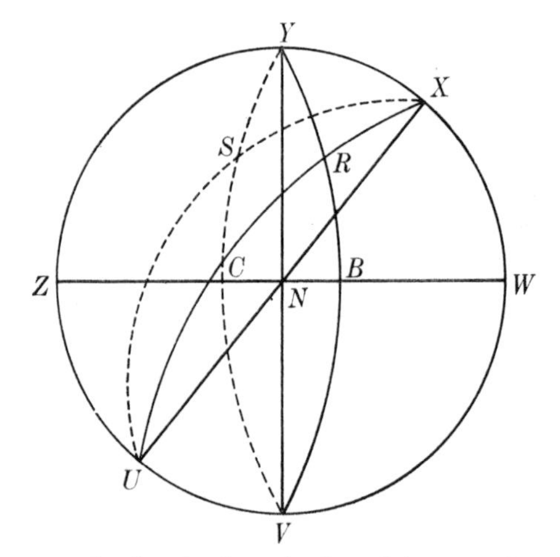

FIG. 4.3. Stereogram showing the determination of the new orientation of a plane *XSU* after a homogeneous gliding from the orientation *XRU* in the direction *WZ*.

both *AD* and the perpendicular to the paper, each point on the sphere is displaced parallel to *AD* by an amount proportional to its distance from *AD*, and the sphere is deformed into the ellipsoid *ACMD*. This ellipsoid has two circular sections that are unchanged in shape and size by the deformation, viz. the section *AOD* on which slipping occurs and the section

OC, which before slipping was the section *OB*. Any other section such as *OP*, which, after deformation becomes *OQ*, changes its shape and also its inclination to *ON*, the normal to the section *AOD*. In the case of a homogeneous deformation by twinning, since the extent of the deformation is a characteristic of the particular crystal, it follows that the sections *OB* and *OC* are unique in being symmetrically disposed on either side of *ON* before and after the deformation.

This last point enables the deformation to be represented conveniently on a stereogram. In Fig. 4.3 the primitive circle *UVWXYZ* represents the plane, and the line *WZ* the direction, of slip. The great circle *VBY* represents the initial position of the circular section *OB* of Fig. 4.2, and *VCY* its position after deformation. (The planes are here represented not by their poles, as is usually the case, but by the great circles to which the planes are parallel.) The new position of any other plane, such as that represented by the great circle *XRU*, may be determined by considering its intersection with the primitive circle and with *VBY*. Any point on the primitive is unmoved by the deformation, while *R* is moved to *S*, the line *RS* being parallel to the diameter *WZ*, so that the new position of *XRU* is given by *XSU*. The behaviour of any other plane is found in a similar way.

4.3. Artificial glide-twinning in calcite

If a blunt knife is pressed into the polar edge of a calcite rhomb, supported in a metal groove, it is deformed as shown in Fig. 4.4. The new face *ABC* is found to be optically plane, proving that the deformation is homogeneous,

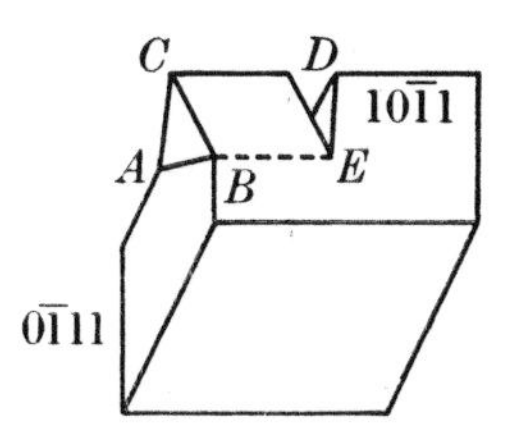

FIG. 4.4. Diagram illustrating Baumhauer's experiment. Gliding parallel to *EB* is caused by pressing a knife edge into the polar edge *CD* of the calcite rhombohedron.

and this face also preserves fixed angular relationships to the other faces of the crystal. The face *BCDE*, and its counterpart behind, remain coplanar with the faces from which they are derived, so that the deformation must take place by slipping on the plane $(01\bar{1}2)$ in a direction parallel to the edge between $(10\bar{1}1)$ and $(\bar{1}101)$, and the plane $(01\bar{1}2)$ is one of the circular sections of the deformation ellipsoid. This point is made clear by a reference to Fig. 4.5(a) which is an ordinary stereogram of calcite rhomb but with the face $(01\bar{1}2)$ in the centre. Fig. 4.5(b) is the corresponding stereogram in which each of the rhombohedral faces is represented by a great circle. The points marked in Fig. 4.5(a) are therefore the poles of the corresponding great circles in Fig. 4.5(b). Since $(10\bar{1}1)$ and $(\bar{1}101)$ remain coplanar with their original positions, they are still represented by the same great circles after twinning and the direction of glide must be *BO*.

The experiment consists in measuring up the angles of the twinned crystal, so that the new position of $(0\bar{1}11)$ may be found, and so that the law of twinning described above may be verified.

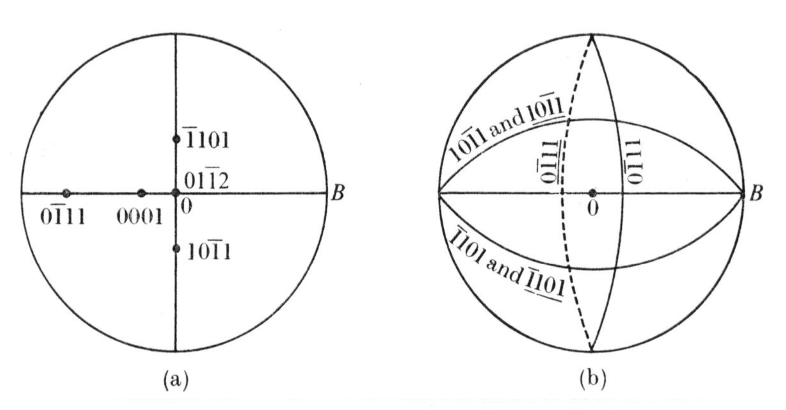

FIG. 4.5. Stereograms, (a) of normals to the faces of calcite, (b) of the same planes represented by great circles.

Experimental details

Some skill, which can only be acquired by practice, is required before calcite rhombs may be twinned satisfactorily. It is therefore advisable to practice twinning on a number of rhombs, and when proficiency has been acquired, a rhomb with optically flat faces is selected. After twinning, a set of measurements of the twinned crystal is made on a goniometer. The usual adjustments should be made to the goniometer before beginning to take the measurements. The crystal is arranged with the zone axis of the zone to be measured approximately vertical and with one pair of faces in the zone parallel to one of the sets of adjusting screws of the goniometer head. The correct orientation of the crystal may then be easily attained. From the measurements two stereograms are constructed, corresponding to Figs. 4.5(a), (b), with a primitive circle of radius 2·5 in; the new face *ABC* (Fig. 4.4) is inserted on each, and the initial and final positions of the second circular section of the deformation ellipsoid are determined.

4.4. The plastic deformation of single crystals of cadmium

4.4.1. *Introduction*

Two processes are associated with the deformation of single crystals of cadmium, namely, gliding and glide-twinning. Gliding is a process in which certain portions of the crystal slide over neighbouring portions along the

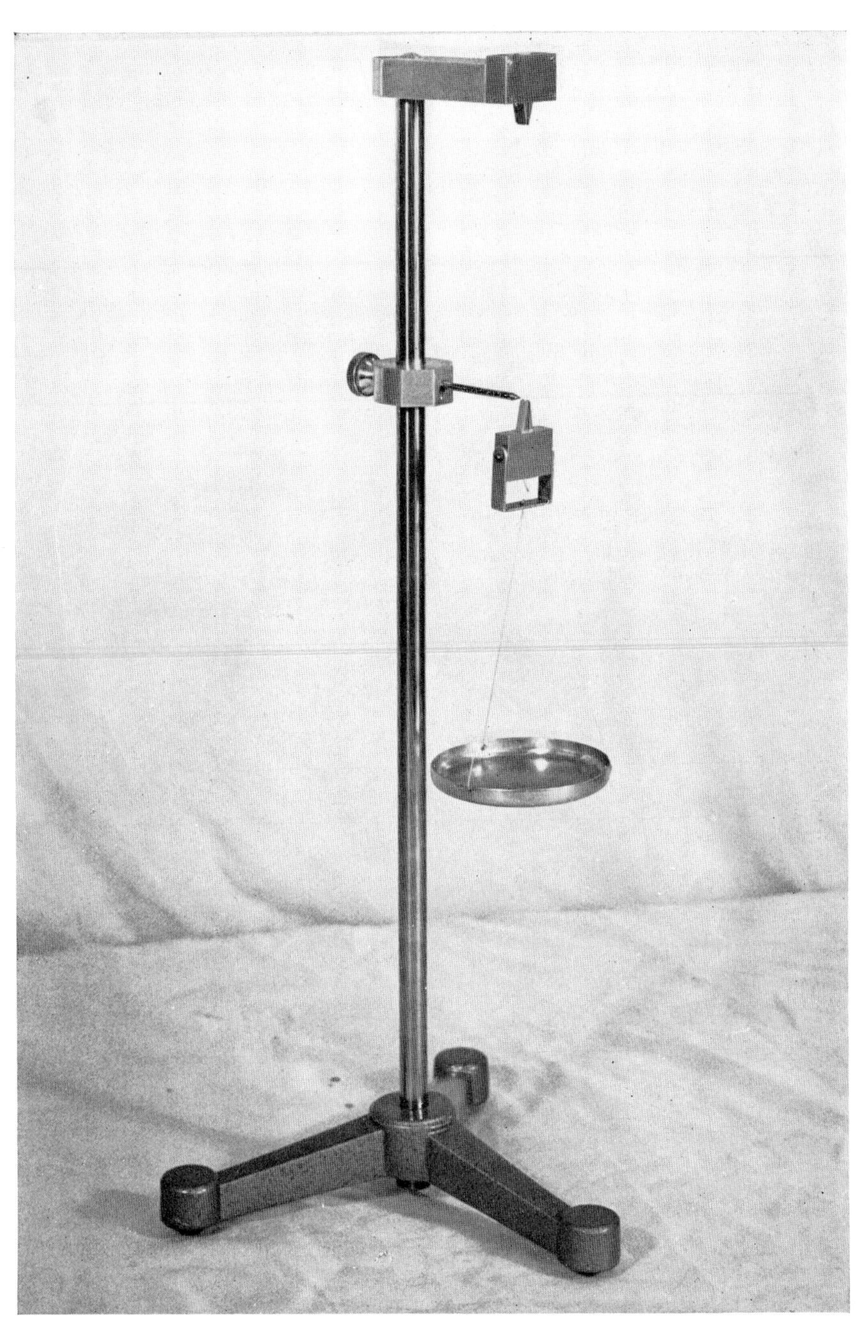

FIG. 4.6. Apparatus used for the stretching of a cadmium single crystal in the form of a wire.

(00.1) plane in the direction of the [100] axis. The glide planes are easily seen where they intersect the surface of the wire. As the wire is stretched dislocations are generated and their mutual interaction offers increasing resistance to further deformation. This process is not quite smooth. Packets of glide planes suddenly yield when the tension reaches a certain value. When the extension has reached a certain amount the glide process stops and another process, namely, glide twinning commences. This is due to the fact that the glide process has brought the glide planes almost parallel to the length of the wire and the component of stress acting to produce gliding has fallen to a small fraction of the applied tension. The effect of the twinning is to produce a further extension of the wire accompanied by a crackling sound. The noise occurs because small volumes of the material suddenly change their orientation and produce a twin having a definite crystallographic orientation with respect of the neighbouring parts of the wire.

4.4.2. *Experimental details*

The wire is held in two clamps which have wedge-shaped jaws so arranged that the greater the tension on the wire the tighter it is held by the clamp (Fig. 4.6). One clamp is carried on an arm attached to a vertical stand. To the other clamp a pan is fixed so that weights may be used to apply tension to the wire. A centimetre scale is used to measure the length of the wire between the points at which it is clamped.

Apply weights to the scale pan gradually and when an extension of 1 cm has been produced note the weights in the pan, remove them and wait 5 min before reloading the pan. When reloading observe carefully the smallest weight which produces further extension of the wire. This weight may be less than the weight removed from the pan. In this case the crystal has partly annealed and some of the work-hardening produced by the deformation has been lost. Now apply more weights. During this part of the loading note the appearance of the surface of the wire, using a lens $\times 8$ approximate magnification. As the loading proceeds listen for the crackling that marks the onset of glide twinning and at the same time note the different surface appearance associated with the portion having a twinned orientation. Plot a curve relating the extension (x-axis) to the load (y-axis) and mark on it the regions corresponding to gliding and to glide-twinning.

The major and minor axes of the final elliptical section should be measured and from the values the area of the cross-section may be calculated. (Area of an ellipse $= \pi ab$, where a and b are the lengths of the semi-axes.) A rough measure of the increase in strength may be obtained by taking the initial strength as the ratio of the load that just produced a visible extension to the area of cross-section, and the final strength as the ratio of the load at the instant of breaking to the final area of cross-section.

EXAMPLE

No representative example can be given because the single crystal wires of cadmium vary so much among themselves. In rare cases a 10-cm long crystal may stretch to 40 cm and the ratio of major to minor axes of the

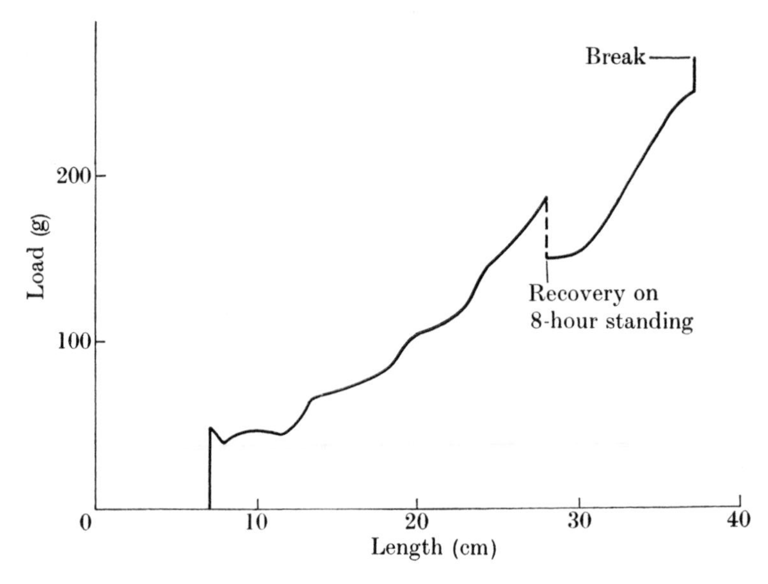

FIG. 4.7. Graph of the load applied in stretching a single crystal of cadmium plotted against its length.

elliptical section may reach a value of 4. The data plotted in Fig. 4.7 relate to a particular crystal. The load was removed for 12 hours before the breaking-point was reached (at a length of 28·5 cm). Annealing had occurred and on again applying loads the crystal increased in length at a lower load than that last applied. After small extensions the occurrence of annealing can be shown after a few minutes release from loading.

5. Dielectric Properties of Crystals

5.1. Introduction

WHEN an insulating crystal is subjected to an electric field a separation of charges occurs. The initial separation occurs on a very local scale; the individual atoms or molecules may become polarized, developing an electric moment determined by the applied field. This polarization is only stable in the best insulators. Usually charges wander through the crystal and eventually build up charged layers on the surfaces of the crystal and these tend to neutralize the effect of the applied field. For this reason nearly all measurements of the behaviour of crystals in an electric field are performed in an alternating field.

The quantity to be measured is known as the dielectric constant, and this may be defined in the following way. A condenser composed of a parallel-sided slab of crystal of thickness t and area A is placed between metal electrodes. The capacity, C, of this system is defined as

$$C = \frac{\mu A}{4\pi t},$$

where μ is the dielectric constant. The area and thickness are here expressed in c.g.s. units. An alternative definition is that

$$C = \frac{\epsilon_0 \epsilon_r A}{t},$$

where ϵ_0 is the permittivity of free space, which in MKS units is equal to $(36\pi 10^9)^{-1}$. A and t are given in m^2 and m respectively. In this definition ϵ_r is a dimensionless quantity, and is known as the relative dielectric constant. C is expressed in farads.

The value of the dielectric constant may depend to some extent on the frequency of the alternating current used. In this case measurements must be performed over a range of frequencies. The frequency must in any case be high enough to avoid the difficulties caused by the drift of charges through the crystal mentioned above.

5.2. Measurement of dielectric constant

A number of methods of measuring the capacity of a condenser are known, including bridge, resonance, and phase-angle measurements. Here we shall deal only with the phase-angle measurements, which are convenient for low-frequency observations. The measurement makes use of the fact that when an a.c. voltage is applied across an ohmic resistance and a capacity in series, the instantaneous voltage across the condenser has a phase in advance of that applied to the system. This difference in phase is expressed by the angle θ and circuit theory shows that

$$\tan\theta = \omega RC = 2\pi fRC,$$

where ω is the angular frequency, f is the frequency in hertz, R the resistance, and C the capacity. In these experiments θ is of the order of 1° and the tangent may be replaced by the angle itself. The principal of the circuit

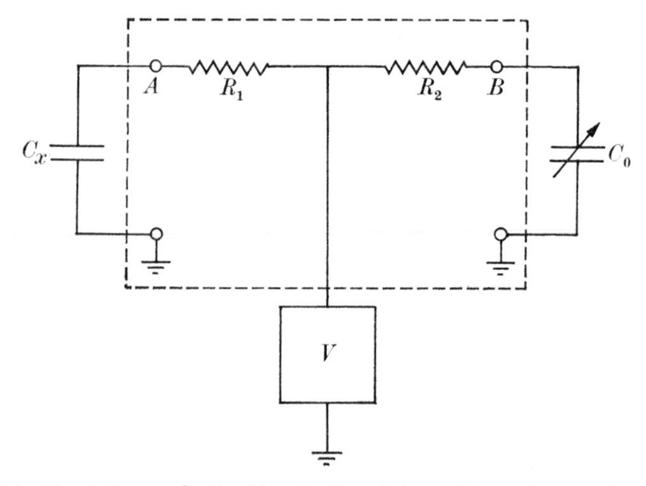

FIG. 5.1. Circuit diagram for the determination of phase difference between the points AB. V represents a variable frequency oscillator.

is shown in Fig. 5.1. The capacity to be determined is marked C_x. A variable condenser is denoted C_0. A variable frequency oscillator supplies a.c. to the two circuits $R_1 C_x$ and $R_2 C_0$ in parallel.

The phase angle at A relative to that of the oscillator is denoted θ_1 and we have

$$\theta_1 = \omega R_1 C_x.$$

The phase angle at B is denoted θ_2 and we have, since $R_1 = R_2$.

$$\theta_2 = \omega R_1 C_0.$$

The difference in the phase angles, which is measured, is denoted θ and

$$\theta = \theta_1 - \theta_2 = \omega R_1(C_x - C_0).$$

In the first example given below C_0 is omitted and the phase angle between the applied voltage and that at point A is measured. In this case

$$\theta = \omega R_1 C_x.$$

The measurement then proceeds in two stages. First with no crystal plate in C_x the capacity of the connecting cable and the much smaller capacity of the parallel plates is measured by finding the value of θ as a function of ω. Then the crystal plate is inserted and again the capacity is found in the same way. The difference of these two measurements gives the capacity when the crystal is between the plates of the condenser.

5.3. Experimental details

The phase meter used in this experiment is a two-stage r.c. coupled difference amplifier followed by a bridge rectifier. The basic circuit is shown in Fig. 5.2. The two input voltages E_1 and E_2 (obtained from points A,

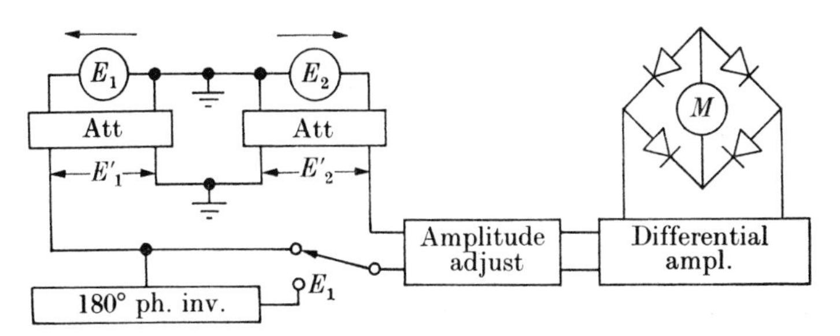

FIG. 5.2. Block diagram of the phase meter used to measure the small capacity due to the crystal. (Reproduced by courtesy of Universal AD-YU Electronics Inc.)

B of Fig. 5.1) are first preamplified and then passed through a step attenuator and a continuously adjustable attenuator. The output of this attenuator is fed to a balanced cathode follower stage which permits adjustment of the input to the differential amplifier. Provision is also made for E_1 to suffer a phase inversion of 180°. This is valuable when the voltages to be compared are nearly 180° apart.

In Fig. 5.2 E_1 and E_2 are two unknown voltages and the arrows indicate their polarities. The meter M reads the output voltage E_0. The relation between E_0, E_1, E_2 and the phase angle θ between E_1 and E_2 can be expressed as follows. If the attenuators are adjusted so that the amplitudes E_1', E_2' are equal but differ in phase by an angle θ we may write

$$E_0 = E_2' - E_1' = E_1'(\cos\theta + j\sin\theta - 1).$$

From this we obtain

$$|E_0| = |E_1'|\sqrt{\{(\cos\theta - 1)^2 + \sin^2\theta\}} = 2|E_1'|\sin\theta/2.$$

The instrument reads 1 V at full scale deflection, i.e. when $E_0 = 1$ V the meter M has a full scale deflection. The range of phase angle corresponding to full scale deflection can be adjusted in the following way.

If E_2 is made equal to zero and E_1' adjusted so that E_0 gives half full-scale deflection, then when E_2' is similarly adjusted so that $E_1' = E_2'$,

$$\frac{|E_0|}{|E_1'|} = 2\sin\theta/2 = 1{\cdot}0,$$

$$\sin\theta/2 = 0{\cdot}5 = 30°,$$

$$\theta = 60°.$$

At full-scale deflexion in this case

$$\frac{|E_0|}{|E_1'|} = 2{\cdot}0$$

and $\theta = 180°$.

If the attenuators are now set so that

$$\frac{|E_0|}{|E_1'|} = 2\sin\theta/2 = 0{\cdot}03478$$

then θ, at full scale deflexion, $=2°$. To achieve this, a certain minimum signal voltage is necessary. The higher the phase angle resolution required the greater must be the available signal strength.

EXAMPLES

(a) Dielectric constant of lead zirconate.

In this example the condenser C_0 was omitted.

$$R_1 = 51\ \text{k}\Omega.$$

Table 5.1 gives the relation between the frequency and the phase angle, with no crystal plate between the condenser plates.

TABLE 5.1

The relation between phase angle and frequency without a crystal dielectric

Frequency f (Hz)	Angle θ degrees	θ/f
100	0·01	0·00010
500	0·105	0·00021
1000	0·215	0·000215
2000	0·43	0·000215
3000	0·65	0·000217

Taking the mean value 0·000216 for θ/f we obtain for the capacity of the coaxial cable and the crystal holder, C_s,

$$C_s = \frac{0{\cdot}000216}{360 \times 51\,000} = 11{\cdot}8 \text{ pF}.$$

A crystal plate having the dimensions 12·46 × 3·19 × 2·54 mm was inserted between the condenser plates and the measurements repeated (see Table 5.2) using a resistance $R_1 = 32$ kΩ.

TABLE 5.2

The relation between phase angle and frequency using the lead zirconate plate

Frequency (Hz)	Phase angle θ (degrees)	θ/f
100	0·233	0·00233
500	1·192	0·002384
1000	2·40	0·00240

TABLE 5.3

Relation between temperature and relative dielectric constant for TGS

Temp. (°C)	Angle (degrees)	Frequency (Hz)	$C = \dfrac{\theta}{360 f R_1}$ pF	$\epsilon_{22} = \dfrac{Ct}{\epsilon_0 A}$
30·5	0·26	1 k	14·2	63
40·0	0·51	1 k	27·8	124
45·0	0·11	100	60	267
46·0	0·146	100	79·5	354
47·0	0·255	100	139	619
47·5	0·45	100	245	1091
48·0	1·18	100	643	2865
48·4	0·86	100	468	2085
48·8	0·62	100	338	1506
50·0	0·335	100	182·5	813
55	0·63	500	68·6	306
60	0·77	1 k	41·9	187

Taking the value for $\theta/f = 0{\cdot}00240$ we obtain for C'_x the value given by

$$C'_x = \frac{0{\cdot}00240}{360 \times 32\,000} = 208 \text{ pF}.$$

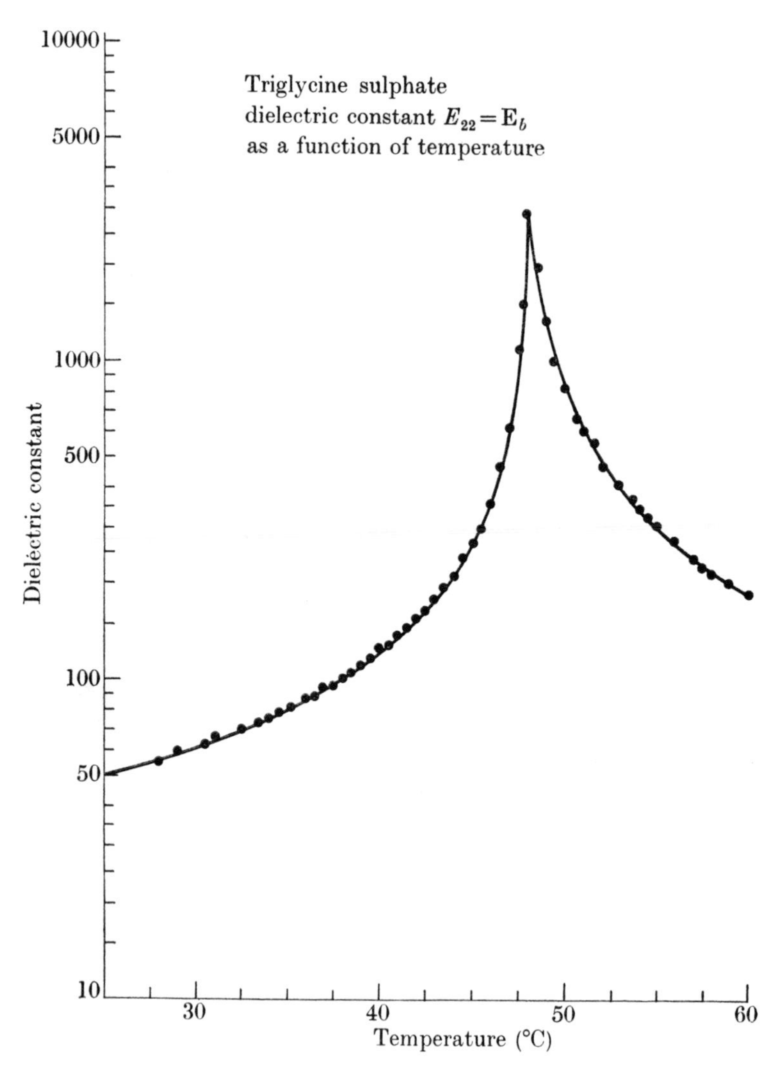

Fig. 5.3. Diagram showing the measured variation of the dielectric constant of TGS as a function of temperature.

Subtracting the value of C_s we obtain the true value of C_x

$$C_x = 208 - 12 = 196 \text{ pF.}$$

The relative dielectric constant is thus

$$\epsilon_r = \frac{C_x . t}{\epsilon_0 . A} = \frac{196 \times 10^{-12} \times 2{\cdot}54 \times 10^{-3} \times 36\pi \times 10^9}{12{\cdot}46 \times 3{\cdot}19 \times 10^{-6}}$$

$$= 1417.$$

(b) Change of dielectric constant of tri-glycine sulphate near its Curie point.

In this experiment the condenser C_0 was used to balance out the capacity of the coaxial cable connecting C_x to the phase meter. The crystal was grown from aqueous solution by the method described in Chapter 10. The plate had its major faces parallel to the (010) cleavage so that the dielectric constant measured is that applying to the *b*-axis, i.e. ϵ_{22}. To keep the phase angle in the measurable range it was necessary to lower the applied frequency near to the transformation temperature, 48°C. The dimensions of the plate were $7{\cdot}21 \times 4{\cdot}12 \times 1{\cdot}17$ mm $R_1 = 51$ kΩ.

These results are plotted in the graph shown in Fig. 5.3 and may be compared with the results of Hoshino *et al.* (1957). Use is made of these results in connection with the study of phase change in Chapter 11.

6. Piezoelectricity

6.1. The constants defining a third-order tensor property and their relation to crystal symmetry

WHEN subjected to mechanical forces some crystals develop surface electrical charges. This phenomenon is known as piezoelectricity. The piezoelectric coefficients of crystals are third-order tensors because they relate a vector and a second-order tensor. The electric moment developed in the crystal, i.e. the charge multiplied by the distance between the charges, may be represented by a vector **p** having components p_i, and the components of stress by a second-order tensor t_{kl}. The most general linear equation that can exist between these quantities is expressed

$$p_i = d_{ikl}\, t_{kl}, \tag{6.1}$$

in which d_{ikl} is a third-order tensor. The quantities d_{ikl} are known as piezoelectric moduli.

If instead of stress components, t_{kl}, we write strain components, r_{kl}, then we have the expression

$$p_i = e_{ikl}\, r_{kl}, \tag{6.2}$$

which is a summary manner of stating that all components of piezoelectric moment are proportional to the strain components that define the deformation of the crystal. The quantities e_{ikl} are known as piezoelectric constants.

One general observation applies to all classes of symmetry in which piezoelectric crystals occur, namely, that those classes must lack a centre of symmetry. This is evident from the fact that the development of opposite charges at the ends of any line passing through the centre of a crystal due to homogeneous deformation is not consistent with there being a centre of symmetry. The variation of the piezoelectric effect with direction in any one crystal can be expressed by means of a representation surface. There are a number of such surfaces, each one corresponding to a particular combination of the type of stress, for example, bending or twisting, and direction of the electric moment relative to the direction of the stress. In Fig. 6.1(a), (b) are shown two sections of one such surface for quartz, namely, the longitudinal piezoelectric surface. Each radius vector of this surface is proportional to the electric moment produced in the direction of the radius vector by a compression of unit magnitude also in the same direction. The piezoelectric modulus determined by the method of section 6.2.1 is proportional to a given radius vector of this surface.

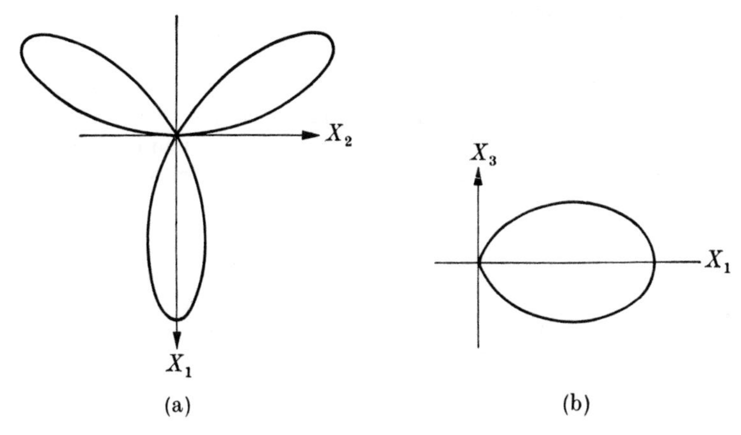

FIG. 6.1. Diagrams of the longitudinal piezoelectric surfaces in quartz, (a) section parallel to (00.1), (b) section parallel to (0.10).

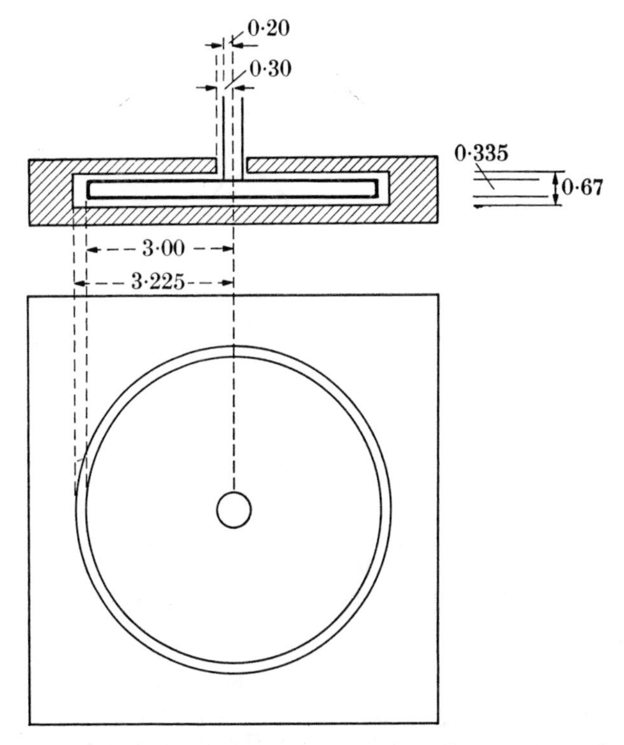

FIG. 6.2. Diagram and dimensions of the standard condenser used to calibrate the readings of the electrometer-triode valve circuit.

6.2. Static method of measuring piezoelectric moduli

6.2.1. *Application of a static compression*

The apparatus consists of a mechanical device for applying pressure to the crystal plate and a means of measuring the electric charge developed on opposite faces of the plate. The crystal is placed on an insulated disc which is connected permanently to the grid of an electrometer triode valve.

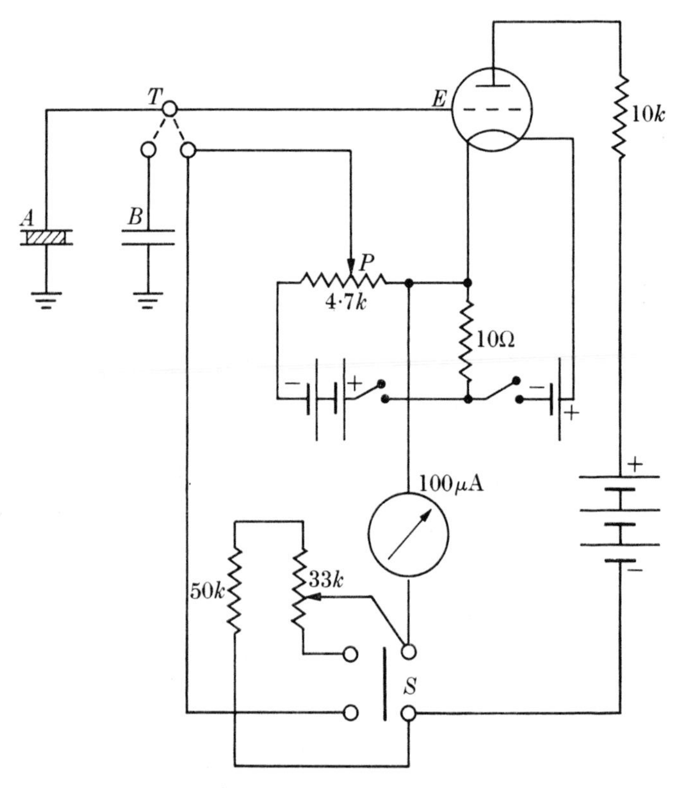

FIG. 6.3. Circuit diagram of the electrometer-triode valve circuit.

Under the insulating rod supporting the crystal is placed a standard condenser. This consists of a disk mounted on insulating spacers within a hollow space, so that the distance between the disk and the walls is well defined. The relevant dimensions are given in Fig. 6.2. This standard condenser may be connected to the grid of the electrometer triode by turning a knob. A further contact operated by the same knob permits the wiper of a 4·7-kΩ potentiometer to be connected to the grid of the valve.

The circuit diagram is given in Fig. 6.3. The crystal *A* forms a condenser which is connected on one side to earth and on the other to the valve *E* which has a highly insulated grid. The standard condenser B may also be joined to E through the three-position switch T. The 4·7-kΩ potentiometer may be joined to E for the purpose of calibrating the anode current in terms of the grid voltage. A change-over switch, S, permits the same meter to be used either for measuring the anode current or the grid voltage.

6.2.2. *Experimental procedure*

The crystal is first inspected to make sure that it is clean and free from moisture. If it appears at all unsatisfactory it should be wiped with cotton wool moistened with absolute alcohol. The crystal is then placed on the insulated plate and the moveable earthed plate allowed to rest on it. A weight of, say, 1 kg is placed on the disk at the top of the rod which is attached to the moveable plate. In this way a known load is applied to the crystal. The supply switch is set to 'on' and the three-position switch, T, is set to 'Bias'. This connects the potentiometer to the valve. The potentiometer is adjusted to give a suitable anode current. The crystal condenser only is now left connected to the grid of E by turning T to 'XTAL'. There should not be a serious drift of the anode current as shown by the meter. If there is, the insulation of the crystal or the valve must be attended to. If the needle is reasonably steady the 1-kg load is removed from the crystal and the change of reading of the meter noted. This observation is repeated with the standard condenser connected to E. Finally, the potentiometer is joined to E, by turning T, and corresponding readings of grid potential and anode current are obtained. These readings permit the conversion of the changes in anode current, obtained by removing the 1-kg load from the crystal, to be converted into changes of potential of the insulated electrode of the crystal.

After making a set of readings with the crystal in one orientation with respect to the insulated plate it should be turned over and a corresponding set of observations obtained. If on removing the load the needle deflected to the right with the plate in one orientation then on turning over the crystal the corresponding deflexion should be to the left.

6.2.3. *Calculation of the piezoelectric modulus*

In compressing the X-cut plate the only stress component is t_{11}, the load per unit area on the major faces. The electric moment per unit volume or charge per unit area is p_1 and the relation between the two is given by

$$p_1 = d_{111} t_{11},$$

where d_{111} is the corresponding piezoelectric modulus. In fact, the charge measured is p_1 multiplied by the area, A, of the plate and the actual load applied is $t_{11} A$. Thus

$$d_{111} = \text{total charge/total load.}$$

The total charge developed is, of course, the same no matter whether the standard condenser is connected or not, but the measured potential is reduced by increasing the capacity. If Q is the total charge due to a given weight W, v_0 the potential measured when the standard condenser is disconnected, and v_1 the corresponding value when it is connected, then

$$Q = v_0 C_0 = v_1(C_0 + C_1),$$

where C_0 is the capacity of condenser A, the wire from the condenser to the valve E and the grid of the valve, and C_1 is the capacity of the standard condenser. From the above equation,

$$\frac{v_0}{v_1} = 1 + \frac{C_1}{C_0},$$

$$\frac{C_1}{C_0} = \frac{v_0 - v_1}{v_1},$$

$$C_0 = C_1 \frac{v_1}{v_0 - v_1},$$

$$Q = v_0 C_0 = C_1 \frac{v_0 v_1}{(v_0 - v_1)},$$

$$d_{111} = \frac{Q}{W} = \frac{C_1}{W} \frac{v_0 v_1}{(v_0 - v_1)}.$$

The values of v_0 and v_1 are measured on the voltmeter, W is the load applied, and C_1 is calculated from the dimensions of the condenser and found to be 34.4 cm (see below).

In compressing an oblique-cut plate we use the relation

$$p_l = d_{111} t'_{33} \sin^3 \theta \cos 3\phi,$$

where p_l is the piezoelectric moment per unit volume in the direction normal to the major faces of the plate, t'_{33} is the stress component normal to the major faces, θ is the angle between the normal to the major faces and the z-axis, and ϕ is the angle between the x-axis and the projection on to the horizontal plane of the normal to the major faces. With this plate we have therefore

$$\frac{Q}{W} = d_{111} \sin^3 \theta \cos 3\phi.$$

6.2.4. *Calculation of the capacity of the standard condenser*

Scale diagrams of the standard condenser are given in Fig. 6.2. For the purpose of this calculation the condenser is regarded as made up of the following parts.

(1) A parallel-plate condenser formed by the upper side of the central plate and the lower surface of the earthed box above it. The diameter of the central plate is assumed to be less than the actual diameter by 0·335 cm so that the edge of the plate can be treated as a hemi-cylinder. On this assumption the capacity is given by

$$\pi(6{\cdot}00 - 0{\cdot}335)^2/(16\pi . 0{\cdot}1675) = 12{\cdot}0 \text{ cm} = 13{\cdot}3 \text{ pF}.$$

(2) A parallel-plate condenser formed by the lower side of the central plate and the upper surface of the earthed box below it. Allowance has to be made for the hole of 6-mm diameter in the lower plate and the rod supporting the central plate. The capacity of the plate is given by

$$\pi(5{\cdot}665^2 - 0{\cdot}3^2)/(16\pi.0{\cdot}1675) = 11{\cdot}9 \text{ cm} = 13{\cdot}2 \text{ pF}.$$

An estimate of the capacity through the fused silica is 0·3 cm = 0·3 pF.

(3) A hemicylindrical condenser of length equal to the circumference of the insulated plate. The inner radius of this condenser can be put equal to half the thickness of the plate, namely, 0·168 cm, and the outer radius has to be given a mean value lying between 0·39 and 0·50 cm. Taking the value 0·45 cm for this radius, the capacity of the condenser is given by

$$2\pi.3{\cdot}00.\frac{1}{2\ln 0{\cdot}45/0{\cdot}168} = 9{\cdot}6 \text{ cm} = 10{\cdot}6 \text{ pF}$$

(4) The contact arm which joins the standard condenser to the insulated plate *B*. An estimate of 0·9 cm is made for this.

The total capacity obtained in this way is 34·4 cm (38·2 pF). The error in this computation may amount to ±2 cm.

EXAMPLE

X-cut plate

Initial reading: 200 μA, corresponding voltage 0·63.

Final readings: 86, 86, 85, 83 μA, mean = 85 μA; corresponding voltage 2·59; change in potential = 1·96 V = v_0.

With standard condenser added:

Initial reading: 200 μA, corresponding to 0·63 V.

Final readings: 157, 157, 157, 157, μA, mean = 157 μA; corresponding to 1·27 V; change in potential = 0·64 V = v_1.

Load used = 1·46 kg.

Remembering that one electrostatic unit of potential is equal to 300 V, we obtain

$$d_{111} = \frac{Q}{W} = \frac{34{\cdot}4}{1{\cdot}46 \times 10^3 \times 981}\frac{1{\cdot}96 \times 0{\cdot}64}{(1{\cdot}96 - 0{\cdot}64)}\frac{1}{300}$$
$$= 7{\cdot}6 \times 10^{-8} \text{ c.g.s. e.s.u.} = 2{\cdot}5 \times 10^{-11} \text{ C/N}.$$

(W. G. Cady, *Piezoelectricity*: $q_{111} = 6{\cdot}9 \times 10^{-8}$ c.g.s. e.s.u.)

The most probable source of error in this experiment is in the calculation of the capacity of the standard condenser.

6.3. Dynamical method

6.3.1. *Introduction*

An alternative to the static measurement described above is the dynamic method in which alternating electric fields produce mechanical standing waves in orientated crystal plates or bars. These measurements are usually carried out on plates fashioned in such a way that only one mode of vibration is present during the observation. If more than one mode of vibration is generated the two interact and the otherwise simple relations between size of crystal bar and the frequency of vibration is disturbed. We shall only discuss here the use of a bar that is long in comparison with its other two dimensions. When such a plate vibrates lengthwise, with a node at the centre and anti-nodes at the ends, the vibration is not seriously affected by coupling to other modes of vibration. Two opposite major faces of such a bar are covered with conducting electrodes. An electrical oscillator is used to generate alternating potentials and these are applied to the electrodes on the plates. The electric field passes through the plate and causes it to expand or contract along its length. A field applied in one direction causes expansion while a field in the opposite direction causes a contraction. The vibrations of the plate are forced by the applied alternating potential and therefore have the same frequency as the applied electric field. The plate is supported at its mid-point so that the ends are free to vibrate. Under these conditions the wavelength of the mechanical wave in the crystal is twice the length of the bar when it vibrates freely in its fundamental mode. Odd order harmonics can be excited but these will not be used in this experiment.

The frequency of the applied electric field may be changed and at a certain frequency the amplitude of vibration of the plate becomes large. This is the phenomenon of resonance that occurs when the frequency of the forcing electric field is near to the natural frequency of longitudinal vibration of the crystal. The velocity of travel of elastic waves through the crystal, v, is given by the relation

$$v = \Lambda f, \tag{6.3}$$

where Λ is the wavelength of the wave in the crystal and f the frequency. In the resonance mode considered here

$$\Lambda = 2l, \tag{6.4}$$

where l is the length of the bar. Thus we have

$$v = 2lf. \tag{6.5}$$

When the bar is cut in an appropriate way only one elastic compliance s is involved in the travelling wave. In this case the relation between the velocity and the compliance is given by

$$v = \sqrt{\left(\frac{1}{\rho s}\right)} \tag{6.6}$$

where ρ is the density. Thus the first result of the measurement is the determination of one value of s. If the bar has its length along the X_1 axis, then the appropriate s is s_{11}. In this case

$$s_{11} = \frac{1}{\rho v^2} = \frac{1}{\rho 4 l^2 f^2}. \tag{6.7}$$

6.3.2. *Forced vibrations near to the resonant frequency*

Mechanical and electrical systems which have a natural frequency of vibration, f_0, and are subjected to a forcing vibration of frequency f behave in similar manners as f rises to f_0 and then passes beyond that value. When $f = f_0$ the system is said to be in resonance with the alternating applied force. The most evident change that occurs when f reaches f_0 is that the amplitude of the vibration produced is much larger than before and that it decreases as f goes beyond f_0. In this experiment we are not, however,

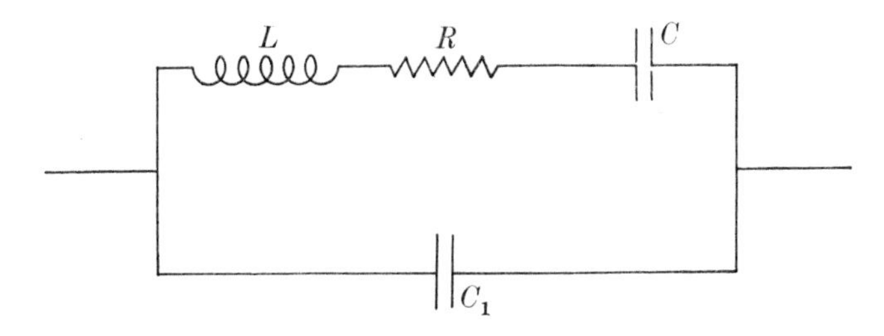

FIG. 6.4. The equivalent electrical circuit of a piezoelectric resonator.

concerned directly with the amplitude of the vibration. The electrical oscillator that produces the alternating potential applied to the electrodes on the plate delivers certain values of current, voltage, and power to the vibrating plate at each frequency. From our measurements we can obtain the voltage across the crystal plate. At frequencies considerably above or below f_0 this voltage has a certain value that changes only slightly with frequency. At a frequency f_s very close to f_0 the impedance becomes very low so that the electrodes on the plates are effectively short-circuited. This is called the series-resonance-frequency. The charge induced on the electrodes by the vibration of the piezoelectric plate almost completely neutralizes the charge transmitted to the plate by the electrical generator when $f = f_s$. At a slightly higher frequency the impedance reaches a maximum. This frequency is called the anti-resonance frequency and is denoted f_p. At this frequency the charge generated by the vibration of the plate has at any moment the same sign as the applied charge and the potential developed across the crystal plate reaches a maximum.

It can be shown that the unloaded piezoelectric resonator behaves in the same way as the circuit shown in Fig. 6.4. Usually the equivalent resistance R is small and the ratio C_1/C, known as the capacitance ratio, is large. When C_1/C is large $(f_p - f_s)/f_s$ is small. It is this quantity which is measured in this experiment. A coefficient of electromechanical coupling, denoted k'_{31}, for a bar elongated along the X'_1 axis and having the electric field applied along the X'_3 axis, can be shown to be given by

$$(k'_{31})^2 = \frac{(d'_{31})^2}{s'_{11}\,\epsilon'_{33}}. \tag{6.8}$$

Where d'_{31} is the piezoelectric strain coefficient for this orientation of the bar, s'_{11} is the elastic compliance involved in vibration along the X'_1 axis and ϵ'_{33} is the dielectric constant for an electric field along the X'_3 axis.

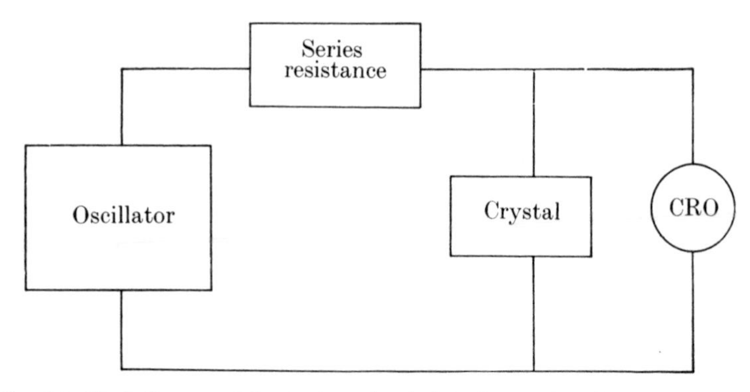

FIG. 6.5. Block diagram of the circuit used to find the frequencies of series and paralle resonances.

The value of k'_{31} is related to Δf, the difference between the parallel and series resonance frequencies, by the expression

$$(k'_{31})^2 = \frac{\pi^2}{4}\cdot\frac{\Delta f}{f_s}. \tag{6.9}$$

Combining eqns. (6.8) and (6.9), we have

$$(d'_{31})^2 = \frac{\pi^2}{4}.s'_{11}.\epsilon'_{33}.\frac{\Delta f}{f_s}. \tag{6.10}$$

Thus measurements of the elastic compliance s'_{11}, of the dielectric constant ϵ'_{33} and the resonance and anti-resonance frequencies are required to find the piezoelectric strain coefficient d'_{31}.

6.3.3. *Experimental procedure*

The apparatus consists of a low frequency oscillator having a maximum frequency of 500 kHz in series with a high resistance and the crystal. The voltage across the crystal is measured by means of a CRO (Fig. 6.5). The

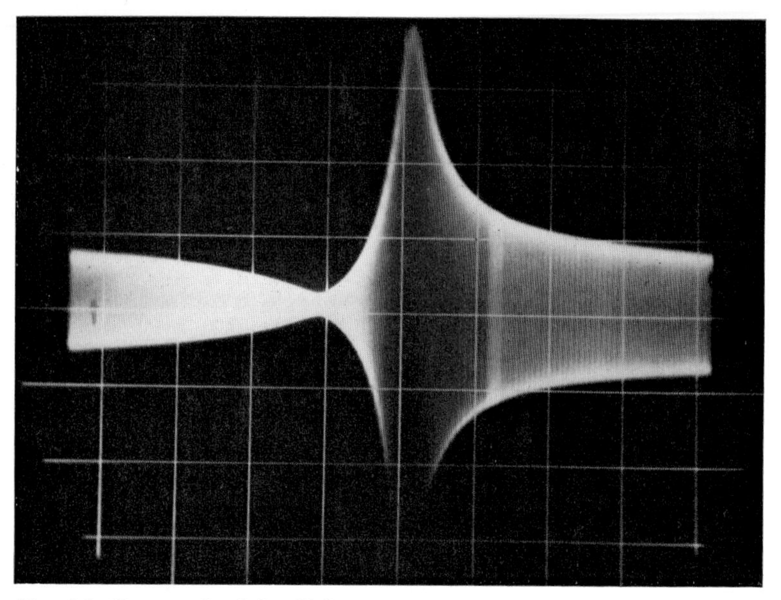

FIG. 6.6. Photograph of the CRO pattern when a sweep through the resonance frequencies is made. The maximum deflection corresponds to the frequency f_p and the minimum to f_s. Range of frequency 24–35 kHz.

crystal bar has plated electrodes and is supported mechanically at its centre. The supports act as electrical connections to the plated electrodes. The best value of the series resistance has to be found by trial and error. Care should be taken not to damage the specimen by too great vibration when it goes into resonance. A ceramic plate will easily sustain a few volts but a thin quartz slab will break under a fraction of 1 V. For this reason the CRO should be used at its highest sensitivity and the voltage applied to the plate kept to a minimum. Covering the whole face with the metallic electrode makes it unlikely that overtones of the fundamental vibration will be excited. As the frequency of the applied potential is raised above that giving resonance a maximum in the voltage across the resonator is observed provided that the relative dielectric constant is high enough. The two frequencies corresponding to series—and to parallel—resonance can be displayed on the CRO by using an electronically or mechanically swept oscillator. Fig. 6.6. shows the result obtained with such a sweep oscillator.

6.3.4. *Example*

TABLE 6.1

Data for lead zirconate ceramic at room temperature

Length (mm)	Breadth (mm)	Thickness (mm)	f_s (kHz)	$f_s \cdot l$	f_p (kc)	Δf (kc)	$\Delta f/f_s$
51·04	6·20	2·02	29·10	1485	30·13	1·03	$3{\cdot}54 \times 10^{-2}$
42	6·2	2·52	35·64	1497	37·06	1·42	$3{\cdot}98 \times 10^{-2}$
27·65	6·31	2·52	54·20	1499	56·21	2·01	$3{\cdot}71 \times 10^{-2}$
23·44	6·22	2·52	63·75	1494	66·21	2·46	$3{\cdot}86 \times 10^{-2}$
Mean				1494			$3{\cdot}773 \times 10^{-2}$

Since

$$v = 2f_s \cdot l,$$

$$= 298\ 800 \text{ cm/s},$$

$$\rho s_{11} = (1/v)^2,$$

$$\rho = 7{\cdot}7 \text{ g/cm}^3,$$

$$s_{11} = 1{\cdot}45 \times 10^{-12} \text{ cm}^2/\text{dyn}$$

$$= 14{\cdot}5 \times 10^{-12} \text{ m}^2/\text{N},$$

$$k_{31}^2 = \frac{\pi^2}{4} \cdot \frac{\Delta f}{f_s},$$

$$k_{31} = \frac{\pi}{2} \sqrt{(3{\cdot}773 \times 10^{-2})} = 0{\cdot}305.$$

The measured value of the relative dielectric constant was 1400 so we write

$$\epsilon_{33} = \frac{1400}{36\pi\, 10^9}.$$

Inserting these values into eqn. (6.8) we get

$$d_{31} = 0{\cdot}305\sqrt{\left(14{\cdot}7 \times 10^{-12} \times \frac{1400}{36\pi\, 10^9}\right)} = 130{\cdot}1 \times 10^{-12} \text{ MKS.}$$

6.4. Detection of the presence of piezoelectricity in a crystal by the method of Giebe and Scheibe

As has been seen in the experiments on quartz resonators, when a crystal goes into oscillation in a circuit it reacts on that circuit, producing a change

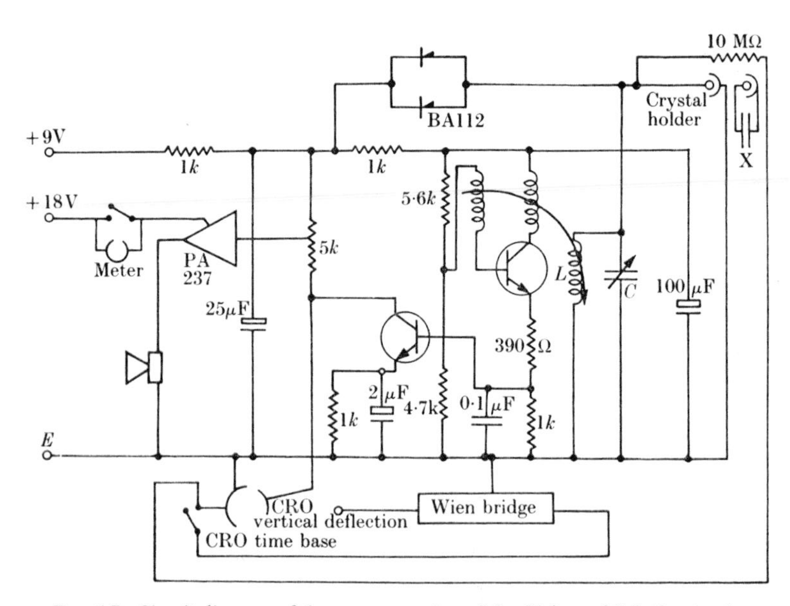

FIG. 6.7. Circuit diagram of the present version of the Giebe and Scheibe circuit.

in d.c. currents. This is the basis of the method, due to Giebe and Scheibe, for finding whether grains of a crystal are piezoelectric. This knowledge is important in crystal-structure determination since the presence or absence of a centre of symmetry determines the choice of space group. The crystal grains are placed between the plates of a small condenser *X* (Fig. 6.7) which is in parallel with a variable condenser *C* and an inductance coil *L*. The variable condenser and the coil form part of a simple oscillatory circuit, the frequency of which changes as the vanes of the variable condenser are rotated. As the frequency sweeps through a certain range of values on turning the vanes, various grains in the condenser X are momentarily excited into oscillation. When this occurs for any one grain there is a sudden change in the d.c. currents which is amplified by the succeeding stages and registered as a click in the loudspeaker or as a peak on the CRO trace if the loudspeaker is replaced by a CRO. The multitude of grains which come into oscillation, each one at a particular frequency, gives rise to a characteristic rustling noise as the vanes of the condenser are rotated. By the occurrence of this noise the presence of piezoelectricity in the grains may be established.

This characteristic sound may be masked by the general noise generated by turning the vanes of the main variable condenser. There is usually a rubbing contact in such condensers and the variation in the contact resistance can give rise, when the piezoelectric effect is weak, to confusing noise from the loudspeaker. To overcome this a pair of variable capacity diodes is placed as shown in Fig. 6.7 in parallel with the variable condenser. A signal of low frequency is applied to this pair of diodes and this causes a variation of the frequency of the oscillatory circuit. The variable condenser and its associated fine-tuning condenser are set to the neighbourhood of one crystal resonance frequency. The variable capacity diodes act as a wobbulator and make the main oscillatory circuit pass through the resonance frequency. Every time such a transit through the resonant frequency occurs the loud-speaker is actuated. The note given out by the loud-speaker has the same, or double, the frequency of the oscillator applied to the diodes. The frequency is doubled if the wobbulator causes the frequency applied to the crystal to pass through the resonant value twice during each cycle of the low-frequency oscillation. It may only pass through it once during each cycle in which case the fundamental note is heard. When the low frequency signal is provided by the same CRO as is used to display the piezoelectric effect the characteristic pattern produced by the resonating crystals is stationary and can be conveniently observed.

7. Pyroelectricity

7.1. The nature of pyroelectricity

CERTAIN non-conducting crystals when heated develop electric charges that are distributed over their surfaces. If a crystal is maintained at a uniform temperature throughout its volume such charges appear only in crystals having a single uniterminal axis. The charges are then developed in equal and opposite amounts at the opposite ends of that axis. This is termed 'true pyroelectricity', and under these conditions the charges developed depend only on the change of temperature, the nature of the crystal, and its cross-sectional area. They are independent of the length. If the crystal is discharged at the high temperature and then allowed to cool, charges of opposite sign appear.

If the temperature of the crystal is not strictly uniform throughout complications arise, for then strains are set up in the crystal which give rise to electric charges that are really piezoelectric in their origin. Such pyroelectricity is termed 'false', and quartz, which possesses three uniterminal diad axes and so cannot display 'true' pyroelectricity, is pyroelectric in this sense. In practice it is extremely difficult to distinguish the two types of pyroelectricity with any certainty, and it is therefore unwise to interpret the presence of pyroelectricity as proving the existence of a single polar axis and it is safer to regard it as indicative only of the absence of a centre of symmetry.

7.2. Methods of detecting pyroelectricity

Pyroelectricity may be conveniently observed in the following ways.

(1) By allowing smoke particles to fall on to the charged crystal. The particles become charged by induction and settle on the charged parts of the crystal whence they grow out in filaments along the lines of force.

(2) By cooling the crystal in liquid air and then suspending it freely in the room. Particles of ice condensed from the atmosphere collect on the charged regions of the crystal and again form filaments.

(3) By detecting the charges directly with an electrometer or similar instrument.

(4) By placing the crystals in a metal spoon and cooling in liquid air. If charges are developed the crystals stick to the spoon and do not fall off when the spoon is inverted.

7.2.1. *Smoke method*

The crystals are placed in an oven at about 150°C until they have attained its temperature. When hot they are removed from the oven and the charges are neutralized by passing the crystals through a spirit flame (which is usually cool enough not to damage the crystal). They are then placed on a glass plate and allowed to cool for a few seconds so that charges are developed again. A piece of magnesium ribbon is ignited and held under an inverted beaker to collect the smoke. The beaker is then placed over the crystals and the smoke allowed to settle. The magnesia grows out into long filaments indicating the lines of force. Drawings should be made, and the experiments repeated so that it may be noticed whether the charges are developed at exactly the same places each time. As the filaments are very fragile it is advisable to make the drawings without disturbing the beaker if possible. A piece of glass treated in the same way as the crystal shows to what extent these filaments form on a non-pyroelectric material.

7.2.2. *Liquid-air method*

Each crystal in turn is suspended by a thread in liquid air. When ebullition has ceased the crystal is removed and allowed to warm up in air. The manner in which the ice particles are deposited on the crystal and the way in which some of these particles are repelled from the crystal after deposition and then travel in space along the lines of force should be noticed. It may be found advisable to heat the crystals in the oven before performing or repeating this experiment, for if the crystals are covered with a film of moisture, when put into the liquid air this will freeze and form a conducting layer of ice which will dissipate the charges produced.

7.2.3. *Flicking electrometer—Gauguin's method*

If the crystal occurs in convenient rod-shaped pieces, as does tourmaline, the following arrangements can be used to show the presence of pyroelectric charges. The crystal T in Fig. 7.1 is mounted in a metal cup fixed to the metal base plate B, and a second metal cup is put on top of the crystal. The electrometer consists of a glass tube K which supports a metal top in which is set an amber insulator carrying a metal rod along its axis. To the lower end of this rod the gold-leaf L is attached and this hangs within 2 mm of an earthed metal plate E attached to B. The cap on the top of the crystal is joined by a wire to L. The crystal is heated by a coil of resistance wire R and the charge developed causes the gold-leaf to be attracted to the plate E. Here it is discharged and so falls back to its original position. This process continues as long as the temperature of the tourmaline is rising—hence the name 'flicking'. When the temperature has reached a steady value the flicking ceases but starts again if the crystal is then allowed to cool. By means of a charged ebonite rod brought near to the top of the electrometer it may be shown that the charges produced on heating are opposite in sign to those that develop as the crystal cools.

The number of flicks is a direct measure of the charge developed, and by discharging continuously the danger of loss of charge by leakage is reduced. It may be shown with this apparatus that the charge developed at either end of an electric axis depends only on the initial and final temperature of the crystal, and not on the rate of heating, provided this rate is great enough to prevent loss by leakage becoming serious. Further, it may be shown that the sign of the charge, but not its magnitude, changes on interchanging the initial and final temperatures; and the charge is proportional to the area of the cross-section of the electric axis but independent of the length of the crystals.

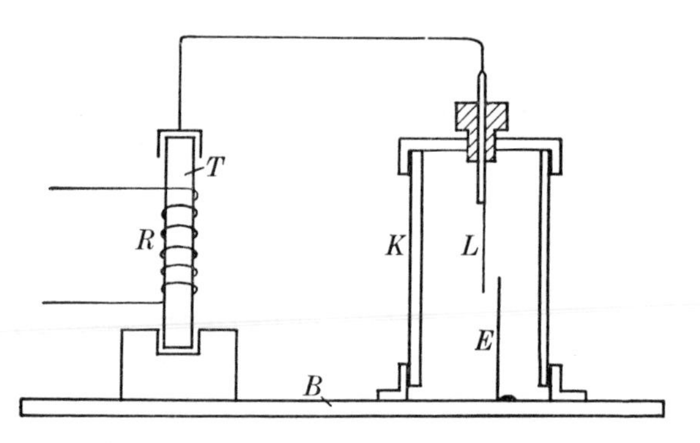

FIG. 7.1. Diagram of Gauguin's flicking electrometer arranged for the demonstration of pyroelectricity in tourmaline.

7.2.4. *Martin's method*

This method differs from the others in that it distinguishes between 'true' and 'false' pyroelectricity, since the experiment is carried out when the temperature of the crystal is uniform throughout. Consequently no effect is observed on quartz, zinc blende, or other piezoelectric crystals that possess more than one electric axis. However, even 'true' pyroelectricity may be derived from the piezoelectric character of the crystal necessarily associated with a unique polar axis.

The apparatus consists of a Dewar flask with an unsilvered strip, containing liquid air, in which is immersed a metal plate *A*, Fig. 7.2, supported on a glass rod which is pivoted at *C*. By means of the long handle it is possible to move *A* a very small distance with certainty. The crystal to be tested, *B*, is suspended by a glass fibre *D*, being tied on with cotton if large enough, or stuck on with mounting paste if very small. The glass fibre is carried on a hook attached to a rotatable head (a cork containing a glass rod) so that the orientation of the crystal with respect to *A* may be varied.

The crystal is lowered gently into the liquid air, and after ebullition has ceased, the metal plate is moved towards the crystal very slowly. A pyroelectric crystal is attracted towards the plate by the induced charge, and clings to the plate as it is withdrawn. Tourmaline and Rochelle salt adhere so strongly that it is difficult to shake them off. In general, the position of the electric axis may be found by rotating the crystal and finding which part adheres most strongly: this is the end of the electric axis. Because of the low temperature the rate of loss of charge is very small. Crystals more or less cubical in shape with sides from 1 to 10 mm may conveniently be employed, or needles and plates 0·1 mm thick provided the electric axis is parallel to the length of the needle or to the plane of the plate.

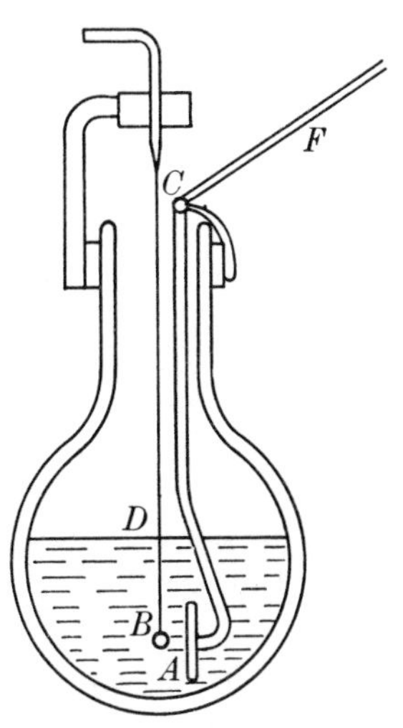

FIG. 7.2. Diagram of Martin's apparatus for the study of pyroelectricity developed by cooling the crystal in liquid air.

A variant on this method which is less sensitive consists in placing some small crystals to be tested in a metal spoon. This spoon is lowered gently into liquid air. If on turning the spoon over the crystal grains adhere to it, then they are pyroelectric. The charge in the crystal induces an equal and opposite charge on the metal spoon and the attraction between these charges is, in this case, sufficient to support the weight of the grains.

7.3. Ballistic method of measuring pyroelectric moments: Maurice's method

A charged crystal A, Fig. 7.3, is suspended in a light cradle by a fibre on the line joining the centres of two oppositely charged spheres B, C. The potentials of the spheres are $+V$ and $-V$, their radii r, the distance apart of their centres is $2l$, and the distance of the centre of the dipole from the centre of the crystal A is d. Under these conditions equilibrium of the crystal is obtained when the deflecting influence of the dipole just balances the restoring couple due to the suspension of the crystal. If the electric moment of A is put equal to M then the couple G, due to the dipole BC, is given by

$$G = M\cos\theta \frac{4rVld}{(d^2 - l^2)^2},$$

where θ is the angle through which the crystal has rotated from its initial position. If the restoring force per unit angular rotation due to the fibre is k, then

$$G = k\theta,$$

from which equation M can be found if k is known and the other quantities are measured. The value of k can be found by placing a bar of known

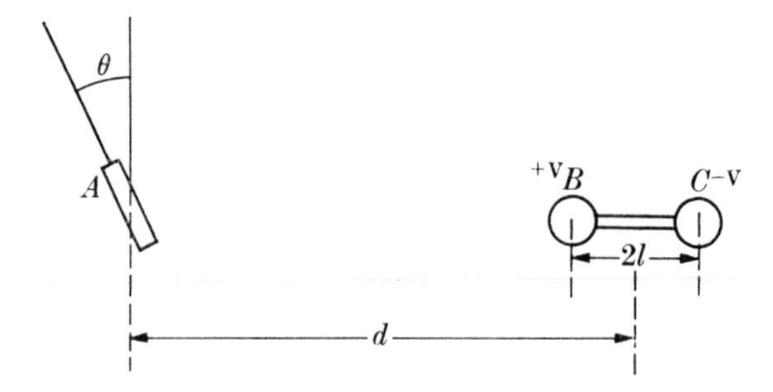

FIG. 7.3. Diagram showing suspended crystal A and electric dipole BC, arranged for the measurement of the electric moment of a pyroelectric crystal.

moment of inertia in the cradle and timing the rate of oscillation. If the period is T and the moment of inertia I, then

$$k = 4\pi^2 I/T^2.$$

EXAMPLE

A tourmaline crystal 2·9 cm long, was heated in a hot-air oven for not less than 15 min and then placed in a light wire cradle of negligible weight. This cradle was suspended by a hair in a large wooden box with a transparent front. An electric dipole, consisting of two brass spheres of 5-mm radius separated by an ebonite rod so that their centres were 4 cm apart, was placed at a convenient distance from the crystal. The line joining the spheres passed through the centre of the crystal. Before applying a potential difference to the brass spheres the length of the crystal was perpendicular to the line joining the spheres. A mirror was attached to the wire cradle and the rotation of the crystal was measured by means of a lamp and scale.

The electric moment Q of the dipole formed by the brass spheres is equal to the product of the charge on each sphere and the distance apart of their centres. The charge on each sphere is the product of its electrostatic capacity, i.e. its radius, and its potential. In this experiment the following conditions applied:

diameter of spheres = 1·0 cm,

distance apart of centres = 4·0 cm,

potentials applied to spheres = +100 V and −100 V respectively.

Hence the charges on the spheres were $+0\cdot5 \times \frac{100}{300}$ and $-0\cdot5 \times \frac{100}{300}$ e.s.u. (300 V = 1 e.s.u. of potential) and

$$Q = 0\cdot5 \times \tfrac{1}{3} \times 4 = 0\cdot667 \text{ c.g.s. e.s.u.}$$

The point midway between the brass spheres was placed 11 cm from the centre of the crystal. This distance is denoted by d.

The strength of the electric field F acting at the crystal may be taken as

$$F = 2Qd/(d^2 - l^2)^2.$$

Thus

$$F = \frac{2 \times 0\cdot667 \times 11\cdot0}{117^2} = 1\cdot07 \times 10^{-3}.$$

The deflexion of the light spot reflected from the mirror due to this electric field was 10 cm. The distance of the mirror from the scale was 100 cm.

The angle of rotation of the crystal was therefore 0·05 rad.

The couple k due to the hair suspension for a rotation of 1 rad was found by replacing the crystal by a brass rod 4·0 cm long, 2 mm diameter and 1·067 g in weight.

The time of oscillation with this brass rod was 10 s.

Using the formula for the moment of inertia, I, namely

$$I = m\left(\frac{l^2}{12} + \frac{r^2}{4}\right)$$

where m is the mass, l the length, and r the radius, we obtain for I the value 1·425 g. cm².

If T is the time of oscillation, then

$$T = 2\pi\sqrt{(I/k)}.$$

Whence we obtain $k = 4\pi^2 I/T^2 = 0\cdot563$ dyn cm/rad. The couple G produced by the electric dipole was therefore

$$0\cdot563 \times 0\cdot05 = 0\cdot0281 \text{ dyn cm.}$$

If the electric dipole of the crystal M was inclined to an electric field F at an angle $(\frac{1}{2}\pi - 0\cdot05)$ rad then the couple acting on it was given by

$$G = M \cos 0\cdot05\,\mathrm{F},$$

or, with the accuracy obtainable in this experiment,

$$G = MF.$$

Hence $M = 0\cdot0281/1\cdot07 \times 10^{-3} = 26\cdot3$ e.s.u.

8. The Hall Effect

8.1. Introduction

In this chapter we have to deal with the effect on moving carriers of electric charge of simultaneously applied electric and magnetic fields. The effect of an electric field alone has been discussed in Chapter 3; the force, F, due to an electric field, E, acting on a charge, q, is the product, qE. When a magnetic field, B, acts on a current, $J = qv$, produced by the charge q moving with a velocity v, the force exerted on the current is at right angles to the plane containing the directions of B and v and of magnitude $qvB \sin \phi$, where ϕ is the angle between the directions of v and B. The so-called right-hand rule relates the direction of the force to the directions of v and B. This may be expressed as a vector equation as follows:

$$\mathbf{F} = q\mathbf{v} \times \mathbf{B}.$$

When the two forces due to the electric and magnetic fields act together the result may be expressed by the equation,

$$\mathbf{F} = q(\mathbf{E} + \mathbf{v} \times \mathbf{B})$$

In this equation the units employed are **F** in newtons, q in coulombs, **E** in volts per metre, **v** in metres per second, **B** in webers per square metre, and **J** in amperes per square metre.

Let us suppose we have positively charged carriers moving in a vacuum. If the motion was due to a positively directed electric field, E_x, along the x-axis, and no magnetic field were applied, the velocity v would be directed in the positive direction of the x-axis. If, in addition, a magnetic field is applied in the positive direction of the y-axis, the force component due to the vector product $\mathbf{v} \times \mathbf{B}$ would be along the positive direction of the z-axis in a right-handed system of reference. This causes the mobile carriers to be deflected and the current density vector, **J**, would have components not parallel to the applied electric field.

In a conductor having defined boundaries, for example a metal wire, both the applied electric field and the current may be in the same direction, in spite of the magnetic field. However, the application of a perpendicular magnetic field B will produce a force component perpendicular both to J and to B. As the mobile carriers cannot move along that direction they must be simultaneously subjected to another equal and oppositely directed force.

Let us separate the force **F** into the parts **F**′ and **F**″ where

$$\mathbf{F}' = q\mathbf{E}$$
$$\mathbf{F}'' = q\mathbf{v} \times \mathbf{B}. \tag{8.1}$$

We shall suppose that the conductor is isotropic and that the mobile (free) carriers are positively charged. Then an electric field directed towards the positive x-direction produces a velocity v_x in the positive direction of x. The application of a positive B_y produces a force F'' directed in the

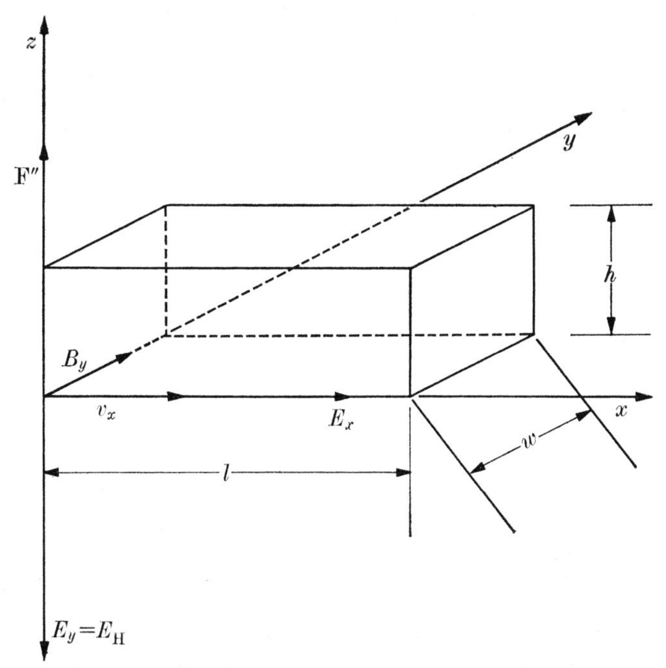

FIG. 8.1. Diagram showing the relative directions of the magnetic field B_y, the electric current, E_x and the Hall electric field E_H.

positive z-direction. This F'' is counterbalanced by a force that can be equated to the action of a field E_y such that

$$E_y = -F'' = -qv_x B_y = q\mathbf{B} \times \mathbf{v} \tag{8.2}$$

This situation is represented in Fig. 8.1. E_y is called the Hall field, E_H, and can be measured with a voltmeter drawing practically no current. The origin of the Hall field can be associated with the crowding of the positive charges on the top face of the sample shown in Fig. 8.1. This face becomes positively charged with respect to the bottom face and the Hall field E_y is due to this charge unbalance.

$$E_H = E_y = \mathbf{B} \times \mathbf{v}. \tag{8.3}$$

If the carriers are negatively charged, for example electrons, then with E_x in the same direction and B in the same direction v is in the negative x-direction and E_H is reversed. This may be seen from eqn (8.3). This means that the top face of the sample becomes negatively charged with respect to the bottom face. Semiconductors often contain carriers of both signs and their effects may neutralize each other.

8.2. The Hall constant

It is convenient to relate the Hall field to the vector **J** rather than the vector **v**. Consider an isotropic medium having both positive and negative carriers. We take p as the number of positive carriers (holes) per unit volume and n as the corresponding concentrations of negative carriers (electrons). When a positively directed electric field, E_x, is applied the following relations hold:

$$J_h = pqv_x, \tag{8.4}$$

$$J_e = nev_{(-x)}, \tag{8.5}$$

where subscripts h and e stand for holes and electrons respectively. In eqns. (8.4) and (8.5) care is taken of the fact that the velocity of the electrons is oppositely directed with respect to that of the holes. We also note that

$$q = -e$$

so that J_h and J_e are both *positive* and *add*. In a material having both types of carriers the total conductivity may be relatively high while the Hall effect is relatively small compared with a material of the same conductivity but having carriers of only one sign. Moreover, in a semiconductor containing only holes and no electrons the Hall field is oppositely directed to that obtained in a metal. These observations are experimental evidence of the existence of positively charged mobile carriers in semiconductors.

Inserting the values of v_x and $v_{(-x)}$ from eqns (8.4) and (8.5) in eqn (8.3) we obtain

$$\mathbf{E}_H \text{ (for holes)} = \frac{1}{pq}\mathbf{B} \times \mathbf{J}_h \tag{8.6}$$

$$\mathbf{E}_H \text{ (for electrons)} = \frac{1}{ne}\mathbf{B} \times \mathbf{J}_e \tag{8.7}$$

The factor $1/pq$ or $1/ne$ is called R, the Hall constant, and it may be seen to be positive for holes and negative for electrons.

The Hall effect is not used to determine the concentration of carriers because this can be easily obtained from the resistivity. It is, however, used to find *mobilities* μ_h, μ_e (velocity under unit electric field) which are different for the two types of carrier. For this reason even when $p = n$ in a semiconductor the Hall field is not zero. The conductivity σ and the resistance ρ are given by

$$\sigma = \frac{1}{\rho} = pq\mu_h \quad \text{(for holes),} \tag{8.8}$$

$$= ne\mu_e \quad \text{(for electrons).} \tag{8.9}$$

The Hall coefficient gives pq or ne (eqns (8.6) and (8.7)) and hence, knowing the resistivity, the mobility may be found.

8.3. Polar and axial tensors

Only in connection with the rotation of the plane of polarized light and the Hall effect does the distinction between polar and axial tensors become

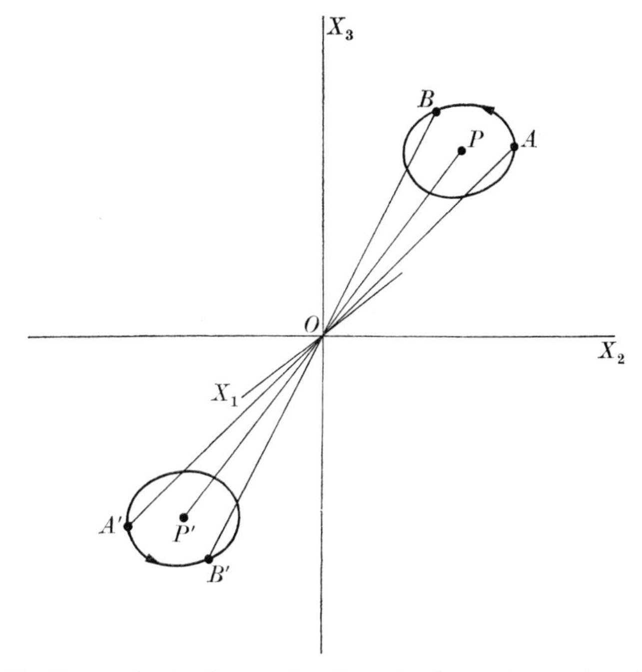

FIG. 8.2. Diagram showing the operation of a centre of symmetry on polar and axial vectors.

important. To make clear the nature of this difference we shall consider a vector OP (Fig. 8.2) which is regarded first as a polar vector and then as an axial vector. The action of a centre of symmetry placed at O on the polar vector OP is to change it to the vector OP' which is of the same length as OP but oppositely directed. When OP is an axial vector the centre of symmetry produces a different result. The vector OP is inverted not to a vector OP' but to a vector $P'O$. This comes about in the following way. An axial vector has associated with its direction a rotation round itself that corresponds to the movement of a nut on a right-handed screw. In Fig. 8.2 a circle is drawn round P and the arrow indicates the sense of rotation round P.

The positive direction of the vector is from O to P. It we take two points A, B on the circle round P and act on them with a centre of symmetry we obtain the points A', B' on a corresponding circle round P'. The sense

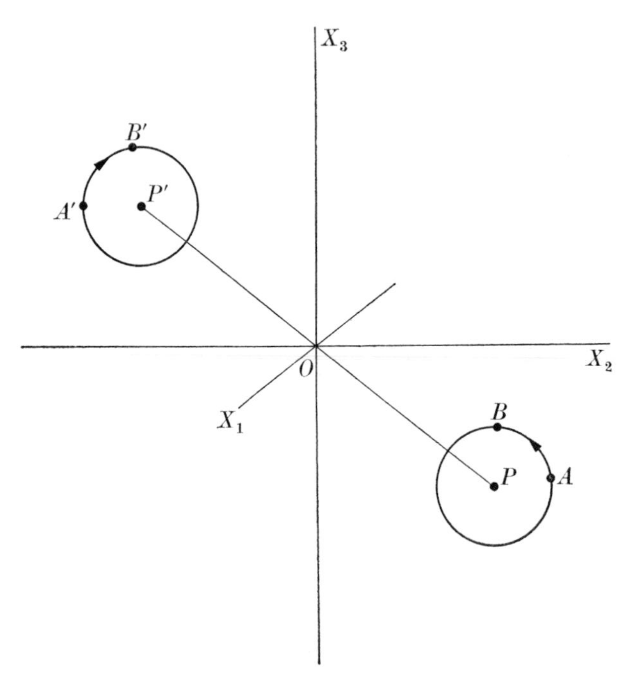

FIG. 8.3. Diagram showing the operation of a diad axis on polar and axial vectors.

of rotation round this circle is from A' to B'. This makes the positive direction of the vector OP' from P' to O, since it must correspond always to a right-handed screw. Thus the vector OP is inverted to one of equal length and having the same direction instead of the opposite direction as with a polar vector. Expressed analytically the components of an axial vector are unchanged by the operation of a centre of symmetry whereas those of a polar vector all change their sign.

A plane of symmetry is like a centre of symmetry in the difference between its action on polar and axial vectors. A rotation axis, however, acts in the same way on both types of vector. This is illustrated by Fig. 8.3, in which X_3 is a twofold axis. Points A', B', are derived from points A, B by rotation about X_3 through 180°. It can be seen that the positive direction of the axial vector OP is towards P and for the vector OP' it is from O to P'. Thus the axial vector is acted upon by the diad axis OX_3 in the same way as is a polar vector PO.

A magnetic field is represented by an axial vector because it is produced by a circular current. This circular current may flow in a coil of wire or it may be due to the rotation of electrons within the atom. In every case a given direction of magnetic field is associated with a sense of rotation which, of course, corresponds to the well-known laws relating electric currents and magnetic fields.

8.4. The Hall constant and crystal symmetry

From eqn (8.6) it is clear that the Hall coefficient R depends on three vector quantities, namely **E**, **J**, and **B**. The quantities **E** and **J** are polar vectors but **B** is an axial vector. R is thus a third-order tensor and the general transformation equation is expressed by the equation

$$R'_{pqr} = \pm c_{pm}\, c_{qn}\, c_{ro}\, R_{mno} \tag{8.10}$$

where the cs are direction cosines.

This differs from the corresponding expression for the piezoelectric tensor by the introduction of the negative sign as well as the positive sign before the terms on the right-hand side. This is due to the axial nature of the vector **B**. The negative sign must be used whenever the symmetry operation involves a centre or a plane of symmetry.

The expression for R derived from eqn (8.6) is

$$E_h = R_{hkl}\, J_k\, B_l.$$

When $k = l$ the magnetic field and the current are directed along the same line and the corresponding Hall effect must be zero. Similarly if $k = 1$ and $l = 2$ the Hall effect is directed downwards along axis 3, whereas if $k = 2$ and $l = 1$ the Hall effect is directed upwards with the same magnitude, i.e.

$$R_{hkl} = -R_{hlk}.$$

These relations impose on the twenty-seven components of the third-order tensor the following restrictions: (1) all diagonal terms are zero, (2) off-diagonal terms occur in pairs of equal magnitude but opposite signs.

Applied to bismuth (class $\bar{3}m$) this gives the scheme:

$$\begin{matrix} 0 & 0 & 0 \\ 0 & 0 & -R_{132} \\ 0 & R_{132} & 0 \end{matrix} \qquad \begin{matrix} 0 & 0 & R_{132} \\ 0 & 0 & 0 \\ -R_{132} & 0 & 0 \end{matrix} \qquad \begin{matrix} 0 & R_{123} & 0 \\ -R_{123} & 0 & 0 \\ 0 & 0 & 0. \end{matrix}$$

There are thus only two coefficients to be determined, namely, R_{132} and R_{123}. R_{132} corresponds to the Hall voltage being developed along axis X_1 due to a current along axis X_3 and a magnetic field along X_2 and R_{123}

corresponds to the Hall voltage developed along axis X_1 due to a current along axis X_2 and a magnetic field along axis X_3. It can be shown that a rectangular plate having its major faces perpendicular to the trigonal axis has a coefficient R_{123} irrespective of the orientation of its length and breadth in the (00·1) plane. Also a rectangular plate having its breadth parallel to the trigonal axis has a Hall coefficient R_{132}, irrespective of the direction of its length.

8.5. Experimental procedure

Using a single crystal of a conductor, or a semiconductor, we usually set the electric field along one of the crystallographic axes and confine the direction of the current to the same axis, so that only one component of the resistivity tensor is involved. The B-vector is also set perpendicular to the J-vector so that the Hall voltage gives a Hall coefficient directly.

Taking the meaning of l, w, and h to be the same as is shown in Fig. 8.1, the actual experimental measurements are

$$I = JA = Jwh, \tag{8.11}$$

$$V_{\mathrm{H}} = E_{\mathrm{H}}.h. \tag{8.12}$$

The potential drop, V, down the length of the sample is given by

$$V = \frac{\rho l}{wh} I, \tag{8.13}$$

where ρ is the resistivity and may be derived from this equation.

The conductivity, σ, is given by

$$\sigma = pq\mu_{\mathrm{h}} \quad \text{or} \quad ne\mu_{\mathrm{e}} = \frac{1}{\rho}, \tag{8.14}$$

depending on whether we have holes or electrons. The combination of the Hall constant R and the conductivity σ gives the mobility μ. From eqn (8.8),

$$\sigma = \frac{1}{\rho} = \frac{1}{R} \cdot \mu$$

or

$$\mu = R\sigma. \tag{8.15}$$

Using the formula

$$E_{\mathrm{H}} = RJB \tag{8.16}$$

and eqns (8.11) and (8.12) we have

$$R = \frac{E_{\mathrm{H}}}{BJ} = \frac{V_{\mathrm{H}}}{h} \cdot \frac{1}{B} \cdot \frac{wh}{I} = \frac{V_{\mathrm{H}} w}{BI}. \tag{8.17}$$

Thus a knowledge of the Hall voltage, the dimension of the sample in the direction of the magnetic field, the magnetic field strength, and the current passed through the specimen gives us the Hall constant. The voltage applied to the sample and the current through it give the resistance and from this and the area of cross-section we obtain σ. Finally, the combination of R and σ yields the mobility μ.

From eqn (8.17) it will be seen that to obtain as large a value of V_H as possible w should be small, B large, and I as large as heating effects permit. Any serious rise in temperature is liable to produce unwanted thermo e.m.f.s.

The measurement of the resistance of the crystal is best carried out as indicated in Fig. 8.4. Two probes a, b are attached on one edge of the sample

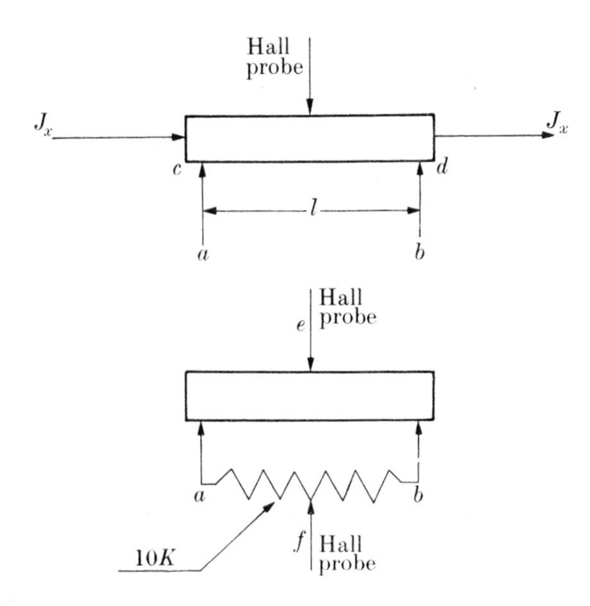

FIG. 8.4. Diagram of the electrodes used in the study of the Hall effect.

and the potential difference between them is measured with an instrument drawing practically no current at all. In this way the effect of contact resistances at the ends or where the probes are attached is practically eliminated. (The probes must not be too near the ends of the specimen. The lines of current flow past the probes must be parallel to the length of the specimen.) From measurements of the length ab and the area of cross-section of the sample normal to I the conductivity σ can be found.

The Hall probes e, f should have no potential difference between them in zero magnetic field. The most convenient way of achieving this is to join the points ab to a potentiometer and then adjust the wiper until, when the current I is flowing, there is no potential difference between e and f.

Then when the field B is applied the voltage read between e and f is due only to the Hall effect.

Certain complications can arise in this measurement. If B is too high the magneto-resistance effect may affect the value of σ. If temperature gradients are produced along the specimen other thermo-electric effects may affect the measurement of R. All these effects can be eliminated by reversing both B and J separately and making a proper linear combination of the results.

EXAMPLES

TABLE 8.1

Conductivity measurement of N-type germanium

$l = 0{\cdot}5$ cm $\quad h = 0{\cdot}2$ cm $\quad w = 1{\cdot}1$ mm

I (mA)	V_{ab} (V)	I/V (mhos)	l/wh (cm^{-1})	$\sigma = \frac{I}{V}\cdot\frac{l}{wh}$ (Ω cm)$^{-1}$
10	1·8	$5{\cdot}56 \times 10^{-3}$		
8	1·4	$5{\cdot}71 \times 10^{-3}$		
6	1·06	$5{\cdot}66 \times 10^{-3}$		
4	0·7	$5{\cdot}71 \times 10^{-3}$		
2	0·36	$5{\cdot}56 \times 10^{-3}$		
Mean		$5{\cdot}640 \times 10^{-3}$	22·7	128×10^{-3}

TABLE 8.2

Hall constant measurement of N-type germanium

B (wb/m^2)	I (mA)	V_H (mV)	V_H/B (MKS)	$R_H = \frac{V_H}{B}\cdot\frac{w}{I}$	$\mu = R\sigma$ (m^2/V s)
+0·87	+10	−240	$-27{\cdot}6 \times 10^{-2}$	$-30{\cdot}4 \times 10^{-3}$	
+0·87	−10	+230	$+26{\cdot}4 \times 10^{-2}$	$-29{\cdot}0 \times 10^{-3}$	
−0·87	+6	+138	$-15{\cdot}9 \times 10^{-2}$	$-29{\cdot}2 \times 10^{-3}$	
−0·87	−6	−133	$+15{\cdot}3 \times 10^{-2}$	$-28{\cdot}1 \times 10^{-3}$	
+1·26	+8	−265	$-21{\cdot}0 \times 10^{-2}$	$-28{\cdot}9 \times 10^{-3}$	
+1·26	−8	+255	$+20{\cdot}2 \times 10^{-2}$	$-27{\cdot}8 \times 10^{-3}$	
−1·26	+2	+66·2	$-5{\cdot}25 \times 10^{-2}$	$-28{\cdot}9 \times 10^{-3}$	
−1·26	−2	−65	$+5{\cdot}16 \times 10^{-2}$	$-28{\cdot}4 \times 10^{-3}$	
Mean				$-28{\cdot}8 \times 10^{-3}$	0·369

Note: This method of operation eliminates the Righi-Leduc and Nernst effects but does not eliminate the Ettinghausen effect.

TABLE 8.3

Conductivity measurement of bismuth (B perpendicular to the trigonal axis)

$l = 5{\cdot}7$ mm $h = 5$ mm $w = 0{\cdot}25$ mm

I (mA)	V_{ab} (mV)	I/V (mhos)	l/wh (cm^{-1})	σ_{33} ((Ω-cm)$^{-1}$)
100	0·78	128		
80	0·6	133		
60	0·45	133		
40	0·3	133		
20	0·15	133		
10	0·08	125		
Mean		132	45·6	6019

TABLE 8.4

Hall constant measurement of bismuth—R_{132}

$B_3' = B_2$ (wb/m^2)	I_2' (mA)	V_H (mV)	V_H/B (MKS)	$R_H = \frac{V_H}{B}\cdot\frac{w}{I}$ (MKS)	$\mu = R\sigma$ (m^2/V s)
+1·02	+100	−0·32	$-3{\cdot}13 \times 10^{-4}$	$-0{\cdot}784 \times 10^{-6}$	
−1·02	+100	0·33	$-3{\cdot}23 \times 10^{-4}$	$-0{\cdot}808 \times 10^{-6}$	
+1·02	−100	0·32	$3{\cdot}13 \times 10^{-4}$	$-0{\cdot}784 \times 10^{-6}$	
−1·02	−100	−0·36	$3{\cdot}52 \times 10^{-4}$	$-0{\cdot}882 \times 10^{-6}$	
Mean				$-0{\cdot}814 \times 10^{-6}$	
+1·02	+60	−0·18	$-1{\cdot}765 \times 10^{-4}$	$-0{\cdot}738 \times 10^{-6}$	
−1·02	+60	0·205	$-2{\cdot}01 \times 10^{-4}$	$-0{\cdot}837 \times 10^{-6}$	
+1·02	−60	0·2	$1{\cdot}96 \times 10^{-4}$	$-0{\cdot}816 \times 10^{-6}$	
−1·02	−60	−0·215	$2{\cdot}11 \times 10^{-4}$	$-0{\cdot}880 \times 10^{-6}$	
Mean				$-0{\cdot}818 \times 10^{-6}$	
−1·02	+20	0·07	$-0{\cdot}687 \times 10^{-4}$	$-0{\cdot}86 \times 10^{-6}$	
+1·02	−20	0·072	$0{\cdot}705 \times 10^{-4}$	$-0{\cdot}88 \times 10^{-6}$	
−1·02	−20	−0·072	$0{\cdot}705 \times 10^{-4}$	$-0{\cdot}88 \times 10^{-6}$	
Mean				$-0{\cdot}87 \times 10^{-6}$	
−1·02	+10	0·034	$-0{\cdot}334 \times 10^{-4}$	$-0{\cdot}835 \times 10^{-6}$	
+1·02	−10	0·037	$0{\cdot}363 \times 10^{-4}$	$-0{\cdot}908 \times 10^{-6}$	
−1·02	−10	−0·034	$0{\cdot}334 \times 10^{-4}$	$-0{\cdot}835 \times 10^{-6}$	
Mean				$-0{\cdot}859 \times 10^{-6}$	
Mean of all values				$-0{\cdot}838 \times 10^{-6}$	0·50

TABLE 8.5

Conductivity measurement of bismuth (B making angles of 45° with X_2 and X_3, J directed along X_1): $R'_{123} = \frac{1}{2}(R_{132} - R_{123})$

$l = 3{\cdot}5$ mm $h = 5{\cdot}8$ mm $w = 0{\cdot}11$ mm

I (mA)	V (mV)	I/V (mhos)	l/wh (cm^{-1})	σ ((Ω-cm)$^{-1}$)
100	0·78	128		
60	0·45	133		
30	0·23	131		
10	0·074	125		
Mean		132	54·9	7247

TABLE 8.6

Hall coefficient measurement of bismuth (B making angles of 45° with X_2 and X_3)

B'_3 (wb/m^2)	I_1 (mA)	V_H (mV)	V_H/B (MKS)	$R = \frac{V_H w}{BI}$ (MKS)
1·02	100	−0·35	$-3{\cdot}43 \times 10^{-4}$	$-0{\cdot}377 \times 10^{-6}$
1·02	60	−0·225	$-2{\cdot}21 \times 10^{-4}$	$-0{\cdot}405 \times 10^{-6}$
1·02	30	−0·11	$-1{\cdot}08 \times 10^{-4}$	$-0{\cdot}396 \times 10^{-6}$
1·02	10	−0·043	$-0{\cdot}42 \times 10^{-4}$	$-0{\cdot}462 \times 10^{-6}$
Mean				$-0{\cdot}410 \times 10^{-6}$

The value of R'_{123} cited above can be obtained by the usual transformation of axes.

From the equation

$$R'_{123} = \tfrac{1}{2}(R_{132} - R_{123})$$

we obtain

$$-0{\cdot}410 \times 10^{-6} = \tfrac{1}{2}(-0{\cdot}838 \times 10^{-6} - R_{123})$$

and hence

$$R_{123} = -0{\cdot}02 \times 10^{-6} \text{ MKS.}$$

9. Elasticity

9.1. The constants defining a fourth-order tensor property and their relation to symmetry

The elastic properties of crystals are determined by the strain experienced by the crystal when subjected to certain types of stress. Expressing this in the most general way possible we write the equations

$$t_{ik} = c_{iklm} r_{lm}, \qquad \text{or} \qquad r_{ik} = s_{iklm} t_{lm}, \tag{9.1}$$

which express the fact that a linear relation exists between each component of stress and all the components of strain and vice versa. The coefficients c_{iklm} of this linear relation are known as elastic stiffnesses or constants while the coefficients s_{iklm} are known as elastic compliances or moduli. In the experiments described in the next two sections compliances are measured because only one stress component is applied and only one type of deformation or strain component is observed. Thus the equation involving s_{iklm} has a solution whereas the left-hand equation cannot be solved since we only observe one r_{lm} out of the six which may be produced by applying the stress. It must be noted that Young's modulus is, in fact, a reciprocal of an s_{iklm}, i.e. a reciprocal of an elastic modulus as defined above.

Te define the elastic properties of a cubic crystal three compliances s_{1111}, s_{1122}, and s_{2323}† must be found. The meaning of these compliances is as follows: s_{1111} is the ratio of the extension per unit length along the X_1 axis to the force per unit area exerted along the same axis. This is, therefore, the reciprocal of Young's modulus for that direction. s_{2323} is the ratio of the change in angle between the X_2 and X_3 axes to the torque applied so as to bring that change about. The compliance s_{12} does not by itself correspond to any usually measured compliance but the compressibility depends on it, being equal to K, where,

$$K = \frac{1}{3(s_{1111} + 2s_{1122})}.$$

It is usually possible to cut plates or bars having such an orientation in relation to the symmetry that some of the compliances may be measured

† These symbols are often abbreviated to s_{11}, s_{22}, and s_{44}. The two suffixes that replace the four are obtained by putting one number for each pair, i.e. for l, m, and for i, k. The substitution is made according to the scheme:

Full symbol	11	22	33	23	31	12
Abbreviated symbol	1	2	3	4	5	6

directly. Other orientations nearly always involve combinations of compliances. There are three fundamental static measurements which may be made, namely, those determining Young's modulus, the rigidity, and the linear or bulk compressibility. In addition a number of different dynamic experiments may be made in which the crystal is caused to vibrate in different ways. From combinations of these different measurements all the elastic coefficients can be obtained.

The variation of any elastic compliance or stiffness with direction in a crystal may be represented by a surface that has the symmetry of the crystal and also a shape that depends on the relative magnitudes of all the elastic compliances but only two are in general use, namely, those corresponding to s_{1111} and s_{2323}. In the experiment described in section 9.2 the radius vector of such a longitudinal extension surface parallel to the diad axis or the triad axis or some intermediate direction is determined.

9.2. Static measurement of Young's modulus by bending a plate

If a bar of rectangular cross-section is mounted between two knife edges *a*, *b*, and two cylinders *c*, *d*, as shown in Fig. 9.1, and a weight is applied to press *a* and *b* downwards relative to *c* and *d* then the part of the bar between *c* and *d* goes into the form of an arc of a circle. Every element

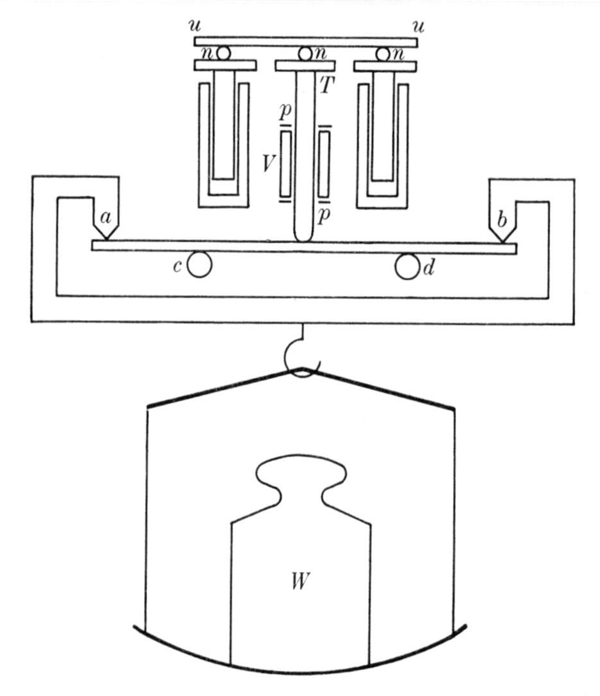

FIG. 9.1. Diagram of an apparatus for bending a crystal plate into an arc of a circle.

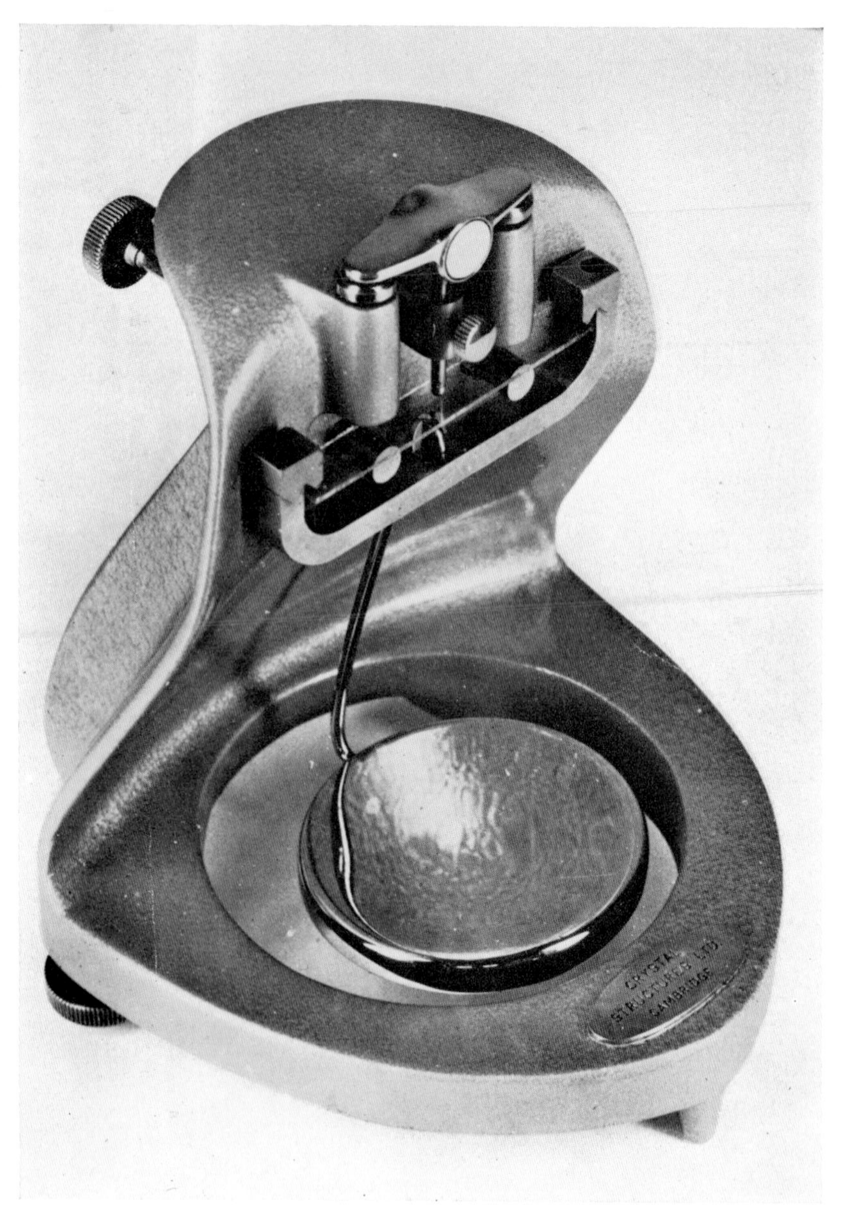

FIG. 9.2. Photograph of the apparatus indicated in Fig. 9.1.

of the crystal above the central plane is stretched parallel to the length of the plate and every element in the lower half is similarly compressed. If the width of the plate is w, its thickness t, and the radius of curvature of the 'neutral' plane passing through its centre is R, then we have

$$\frac{1}{R}=\frac{12M}{wt^3E},$$

where M is the bending moment acting across the section, and E is Young's modulus for the direction of the length of the plate. If x is the upward displacement of the crystal and $2a$, $2b$ the distances between the cylinders and the knife edges respectively, then

$$\frac{1}{R}=\frac{2x}{a^2}.$$

If the mass put in the scale-pan is W, then

$$M=\tfrac{1}{2}Wg(a-b).$$

Finally,

$$E=\frac{6Ma^2}{xwt^3}.$$

Experimental details

The crystalline bars used in these experiments are of quartz and they are mounted as shown in Figs. 9.1 and 9.2. A weight W is placed in the scale-pan to apply the bending moment. It is best to place corrugated cardboard or rubber sheet in the scale-pan so as to avoid giving sudden jars when putting on the load. The radius of curvature of the portion between the cylinders c, d is measured by finding the rise of the centre of the plate. The movement of the centre of the plate is transmitted to a vertical rod at the top of which is a table T. The rod is supported in a tube V which in turn is carried on two flat springs p, p, which are in turn fixed to the back-plate of the instrument. The springs only permit vertical movement of the rod T. A small glass plate is fixed to the top of the table T. Two other glass plates are similarly arranged, one on each side of T. They are carried on vertical rods which slide up and down in vertical holes drilled in the base casting. Adjusting screws at the back of the instrument can raise and lower these vertical rods so that all three glass plates can be brought into the same plane. A balance-unit u, u consisting of three steel balls n, n, n, almost in a straight line, rests on the three pieces of glass. Mirrors mounted vertically are attached to both sides of the balance-unit. (Though only one is necessary for measurement, the second is attached to make the unit symmetrical.) These mirrors form part of an optical lever by which the rotation of the balance-unit may be measured. The middle steel ball should be displaced by about 0·5 mm to one side of the line joining the outer balls and rests on Table T. The outer balls rest on the other pieces of glass. When the three glass plates are co-planar and the

8

balance unit is resting on them the plane through the balls is horizontal and perpendicular to the plane of the mirror. If now an upward bend of the crystal raises the table *T*, the normal to the mirror moves upwards or downwards according to the orientation of the balance-unit. A conventional lamp and scale is used to record the movement of the mirror. The balance-unit must be nearly unstable in order to afford the necessary optical magnification of the deflexion of the centre of the crystal plate. If one millimetre deflexion on the scale corresponds to rather less than a wavelength of blue light, the sensitivity will be adequate. The table *T* should always be adjusted relative to the tube *V*, so that a slight pressure is exerted by springs *p*, *p* on the crystal plate when measurements are being made.

The load is applied to the crystal by means of the two knife edges *a*, *b*, forming part of a stirrup (Fig. 9.1).

The crystal plate is seldom cut perfectly plane and small differences are observed according to which way it is placed in the instrument. There are four different ways of putting in the plate and the accuracy is improved if all of them are used.

The calibration of the deflexion is best done by using a piece of nickel foil such as is employed for a β-filter in X-ray crystallography. The foil may be cut into an area of some 50 cm^2 and weighed in order to find its thickness accurately. A small piece of the foil is placed on top of the glass on table *T* and the deflexion produced when the balance-unit is replaced is noted. The foil is removed, the balance-unit replaced, and the whole measurement repeated. It is necessary for this calibration that all the three pieces of glass should be co-planar. The balance-unit is capable of small movements towards or away from the lamp and when the pieces of glass are co-planar or, at least, are in parallel planes, there is no change of deflexion produced by such displacements of the balance-unit. During the measurement of the deflexion of the crystal plate this parallelism of the glass plates is less important because the balance-unit is left untouched. However, when using the foil to calibrate the scale, the balance-unit must necessarily be set down in slightly different positions on successive occasions. It is essential that such slight differences should not alter the deflexion of the light beam.

EXAMPLE

Two plates were used having the following orientation and dimensions.

TABLE 9.1

Dimensions and orientation of quartz plates

	Length parallel to	Thickness parallel to	Thickness (mean cm)	Width (mean cm)
Y-cut	*Z*	*Y*	0·1634	0·997
Z-cut	*X*	*Z*	0·1583	0·998

The thickness was measured at nine points, and the width at three. The load used was 258 g, giving a force of $2{\cdot}53 \times 10^5$ dyn.

Distance between inner cylinders $= 4{\cdot}50$ cm.

Distance between outer knife edges $= 7{\cdot}50$ cm.

Hence bending moment $M = \dfrac{2{\cdot}53}{2} \times \dfrac{(7{\cdot}50 - 4{\cdot}50)}{2} \times 10^5$

$= 1{\cdot}898 \times 10^5$ dyn cm.

Calibration of scale

Dimensions of rectangular nickel foil $= 8{\cdot}01 \times 7{\cdot}88$ cm (area $= 63{\cdot}1$ cm^2).

Weight of foil $= 0{\cdot}9185$ g.

Density of nickel $= 8{\cdot}9$.

Volume of foil $= 0{\cdot}9185/8{\cdot}9 = 0{\cdot}1032$ cm^3.

Thickness of foil $= 0{\cdot}1032/63{\cdot}1 = 0{\cdot}00163(5)$ cm.

TABLE 9.2

Calibration of instrument by using a foil of known thickness

Scale readings		
With foil (mm)	Without foil (mm)	Deflexion (mm)
−4·3	44·1	48·4
−1·8	46·2	48·0
0·0	47·6	47·6
2·0	48·5	46·5
1·0	48·2	47·2
−1·8	46·7	48·5
		Mean 47·7

Movement of central ball corresponding to 1 mm deflexion on scale

$$= 0{\cdot}001635/47{\cdot}7 = 3{\cdot}43 \times 10^{-5} \text{ cm};$$

if the scale reading is q mm then

$$x = q \,.\, 3{\cdot}43 \times 10^{-5}.$$

Hence

$$E = \frac{6Ma^2}{3{\cdot}43 \times 10^{-5}} \frac{1}{qwt^3}$$

$$= \frac{6 \times 1{\cdot}898 \times 10^5 \times 5{\cdot}06}{3{\cdot}43 \times 10^{-5}} \frac{1}{qwt^3} = \frac{1{\cdot}680 \times 10^{11}}{qwt^3}.$$

TABLE 9.3

Deflexion of light-spot on bending a Y-cut plate

Orientation of plate	No load (mm)	Load 258 g	Deflexion (mm)
1	37·0	−2·7	39·7
	36·3	−3·2	39·5
	35·9	−3·7	39·6
2	38·6	−0·8	39·4
	38·2	−0·9	39·1
	38·0	−1·1	39·1
3	36·8	−2·2	39·0
	36·5	−2·1	38·6
	36·4	−2·3	38·7
4	43·1	3·9	39·2
	43·1	4·0	39·1
	43·1	4·0	39·1
			39·2 Mean

$$E = \frac{1{\cdot}680 \times 10^{11}}{39{\cdot}2 \times 0{\cdot}997 \times 0{\cdot}1634^3} = 9{\cdot}9 \times 10^{11},$$

hence

$$s_{33} = \frac{1}{E} = 10{\cdot}1 \times 10^{-13}\,\text{cm}^2\,\text{dyn}^{-1}.$$

Z-cut plate

Observations in the four orientations of the plate gave a mean-scale deflexion = 54·1 mm.

$$E = \frac{1{\cdot}680 \times 10^{11}}{54{\cdot}1 \times 0{\cdot}998 \times 0{\cdot}1583^3} = 7{\cdot}8 \times 10^{11}$$

hence

$$s_{11} = \frac{1}{E} = 12{\cdot}8 \times 10^{-13}\,\text{cm}^2\,\text{dyn}^{-1}.$$

(Mason, W. P. *Piezoelectric crystals and their application to ultrasonics*: $s_{11} = 12{\cdot}79 \times 10^{-13}$, $s_{33} = 9{\cdot}56 \times 10^{-13}$ cm^2 dyn^{-1}.)

9.3. Dynamic measurement of one compliance for a crystal of KDP (KH_2PO_4)

The method of exciting a piezoelectric plate into resonance with an applied alternating electric field has been described in Chapter 6. Provided the crystal bar is long in comparison with its other dimensions, the frequency at which resonance occurs with a longitudinal vibration, involving an alternating increase and decrease in the length of the bar, is independent of other modes of vibration. If the length of the bar is along the axis X_1 the relevant compliance is s_{11} and the relation between the velocity v, the density ρ, and the compliance s_{11} is (see eqn. (6.6))

$$v = \sqrt{\left(\frac{1}{\rho s_{11}}\right)}.$$

When the bar vibrates in its fundamental mode the length of elastic wave is twice the length of the bar since the ends are antinodes and the centre is a node. This leads to the result (see eqn. (6.7))

$$s_{11} = \frac{1}{\rho 4 l^2 f^2}$$

where l is the length of the bar and f the frequency of vibration.

If the bar is cut parallel to the X_3 axis the compliance involved is s_{33}. In directions of a more general character several compliances may be involved. It is usually necessary to prepare several bars cut in different directions in order to find the various values of the s_{ik}'s. The values of the s'_{11} corresponding to the compliance for a bar cut in a general direction can be found from the standard works of reference, for example W. G. Cady, *Piezoelectricity*, Chapter 4. In the example that follows bars having their lengths in the (001) plane of a tetragonal crystal have been chosen. In this case s_{12} and s_{66} are involved as well as s_{11}.

EXAMPLE

Z – cut KDP (KH_2PO_4) – length inclined at $\theta°$ to X_1 – axis at room temperature.

Length (mm)	Breadth (mm)	Thickness (mm)	f (kc)	$f.l$	Orientation θ (degrees)	$\cos^2\theta . \sin^2\theta$
12·44	4·85	1·02	119·54	1487·08	45	0·25
19·59	3·23	1·08	75·27	1474·54	45	0·25
12·86	3·07	1	115·6	1486·62	45	0·25
Mean				1482·75		
15·58	4·18	2·40	138·58	2159·08	15	0·062

Volume 19 cm³ Mass 42·53 g

For the 45° orientation:

$$v = 148\,275 \times 2 = 296\,550 \text{ cm/s},$$

$$\rho = 42{\cdot}53/19 = 2{\cdot}24 \text{ g/cm}^3,$$

$$s'_{11}(45°) = 1/\rho v^2 = 5{\cdot}08 \times 10^{-12} \text{ cm}^2/\text{dyn}.$$

For the 15° orientation:

$$v = 215\,908 \times 2 = 431\,816 \text{ cm/s},$$

$$s'_{11}(15°) = 1/2{\cdot}24 \times 431\,816^2 = 2{\cdot}39 \times 10^{-12} \text{ cm}^2/\text{dyn}.$$

It can be shown using the usual transformations that

$$s'_{11}(\theta) = s_{11} + \cos^2\theta . \sin^2\theta(-2s_{11} + 2s_{12} + s_{66})$$

where $s'_{11}(\theta)$ is the measured value of s for the bar having its length in the (001) plane and inclined at an angle θ to X_1.

This may be written

$$s'_{11} = a + kb.$$

Thus we have the equations

$$s'_{11}(45°) = 5{\cdot}08 \times 10^{-12} = a + 0{\cdot}25b,$$

$$s'_{11}(15°) = 2{\cdot}39 \times 10^{-12} = a + 0{\cdot}062b.$$

Solving for a and b we obtain

$$a = s_{11} = s_{22} = 1{\cdot}51 \times 10^{-12} \text{ cm}^2/\text{dyn},$$

$$b = 2s_{12} + s_{66} - 2s_{11} = 14{\cdot}31 \times 10^{-12}.$$

and hence

$$2s_{12} + s_{66} = 17{\cdot}33 \times 10^{-12} \text{ cm}^2/\text{dyn}$$

using only bars of this kind and with this method it is not possible to find s_{12} and s_{66} separately.

10. Crystal Growing

10.1. Introduction

THERE are several ways in which single crystals can be grown, two of which will be described here, namely, growth from solution and growth from a melt. Under each of these headings there are several different methods and one particular method is dealt with in each case. The solutions considered here are all aqueous and the growth is caused by the cooling of a saturated solution. The growth from the melt concerns cadmium in the form of thin wires suitable for the study of its plastic properties.

10.2. Growth from solution

All the crystals discussed in this section are ionic salts and are highly soluble in water. The purity of the solute is an important factor in determining the quality of the crystals obtained. The information on the chemical composition of the salts used is often inadequate in indicating whether disturbing impurities are present or not. Further, some impurities affect the result when present in such small amounts that they cannot be detected chemically. It might be thought necessary to use only materials of the highest purity. This is not so and sometimes impurities must be added. For instance, to obtain good results with NaCl it is necessary to have lead present and cheap common salt contains enough of this element, whereas the analytically pure material does not. In some cases, though not in those discussed below, the pH of the solution is a critical factor. The additives necessary for good growth in the case of the crystals discussed here are given in Table 10.1.

TABLE 10.1

Additives for assisting growth

Abbreviation	Chemical formula	Additive
TGS	$(CH_2NH_2COOH)_3H_2SO_4$	P_2O_5
ADP	$NH_4H_2PO_4$	Fe^{3+}, Cr, Al, Sn
KDP	KH_2PO_4	Borax
	$NaClO_3$	
OAM	$(NH_4)_2C_2O_4 . H_2O$	Glycerine

These additives may not in every case prove necessary. For instance, the authors have obtained perfectly shaped KDP crystals without borax, for some years, during which the same firm supplied the KDP; but when the source of supply was changed borax was found to be necessary. The amount of additive is generally a few per cent but this should be determined by trial and error.

10.2.1. *An apparatus for growing crystals*

Figure 10.1 is a diagram of the apparatus required for growing crystals by the most easily controlled method, namely, that of lowering the temperature of a saturated solution. The solution is contained in a beaker,

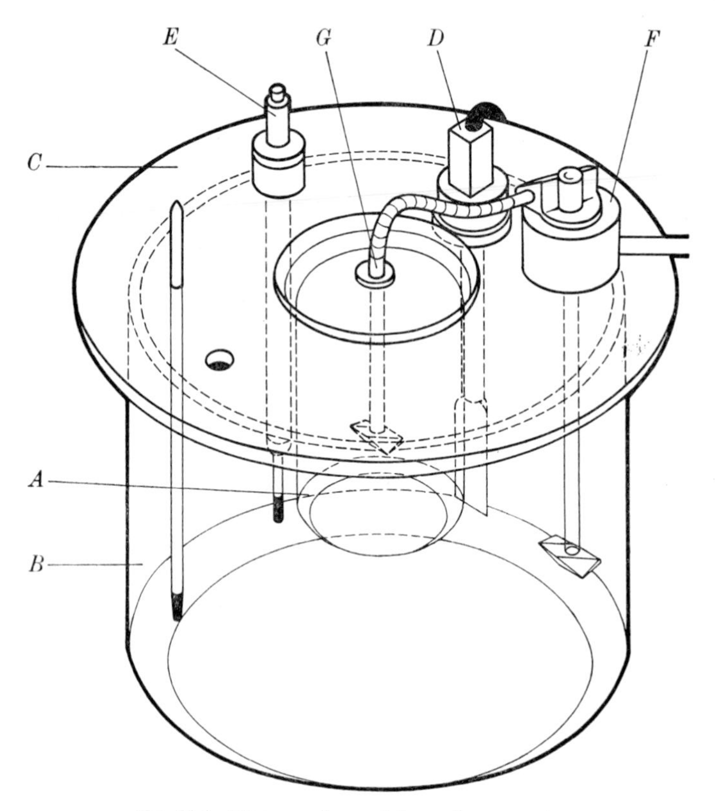

FIG. 10.1. Diagram of a crystal-growing apparatus.

A, holding 2 l, which is mounted coaxially with an outer vessel, *B*, holding 35 l. The outer vessel contains water at a temperature that is controlled by a thermo-regulator. It is covered by a bakelite lid, *C*, approximately 0·5 inches thick, having a central hold just large enough to admit the inner

FIG. 10.2. Photograph of the apparatus indicated in Fig. 10.1.

vessel, *A*. Bakelite is used because of its mechanical stability at the temperatures used. The beaker has an upper rim that extends outwards and on which it is supported. To prevent the beaker from being lifted by the water in the outer vessel it is held down by a perspex cover screwed to the bakelite cover by four wing-nuts (see Fig. 10.2). This is necessary because the level of the water in the outer vessel is often higher than that of the solution in the beaker.

The temperature controller operates satisfactorily when the ambient temperature is maintained within the range 26 ± 1°C. If the temperature of the room rises above 27°C the whole experiment is liable to be ruined.

The water in the outer vessel is heated by an immersion heater of 250–300 W. This is controlled by a mercury-in-glass differential thermo-regulator. graduated from 0°C to 70°C (*E*, Fig. 10.1). The fine adjustment of this thermostat provides about 0·3°C per turn and can be driven by a synchronous clock motor at a few turns per day. A better arrangement is that in which a miniature d.c.-motor drives the thermo-regulator, the voltage applied to the d.c. motor being controlled from a regulated power supply. The maximum current that may be passed through the thermo-regulator is 15 mA. A relay or controlled diode is required suitable for operation at this current. Other types of thermo-regulator, e.g. those based on bimetallic strips, often stick and spoil the growth.

The water in the waterjacket is stirred vigorously by an agitator operated by a motor, *F* (115 V. 0·5 A). The motor has two output shafts rotating at 1700 rev/min and 350 rev/min respectively. The second shaft is used to stir the solution when required. The solution stirrer is connected by a length of rubber tube to the motor shaft. The stirrer has a glass shaft and teflon propeller; metals must be avoided in these solutions because of corrosion.

10.2.2. *Experimental procedure*

(a) *Preparation of the solution*

It is unwise to rely on published solubility curves. During the making-up of the solutions at 60–65°C evaporation occurs and this renders precise measurements of weights of solute and solvent not very useful. It is better to avoid supersaturation by providing a solution at a slightly higher temperature than required and then allowing it to cool slowly, until spontaneous crystallization occurs. The solution is then saturated, or slightly supersaturated.

The quantities given in Table 10.2 are approximate and should provide solutions that are close to being saturated, so that only small corrections are needed. A hot plate equipped with a magnetic stirrer is the most convenient means of getting the solute into solution. If the hot plate is thermostatically controlled it is possible to arrange with a usual volume of solution that the temperature rises only slowly through the required point.

For all the salts except TGS it is best to raise the temperature of the

TABLE 10.2

Relative quantities of solution and solvent (to be used at temperatures 5°–10°C lower than given in the table)

Crystal	Weight required (g)	Distilled water (cm^3)	Temperature (°C)
TGS	182 g glycerine 43 cc H_2SO_4	500	60
KDP	350	750	60
ADP	500	750	60
$NaClO_3$	113	100	70
OAM	150	1000	55

water to the required temperature before adding the salt because the process of solution lowers the temperature of the solvent. In preparing the solution for TGS the sulphuric acid is added last in very small quantities at a time as the temperature tends to rise. Stirring is continued until all the salt is dissolved. The temperature may be raised some 5°C except with OAM which at 55°C is rather near its decomposition temperature.

As soon as all the salt is dissolved it is filtered as quickly as possible using a filter pump. Part of the solution is transferred to a small beaker for seed growth at room temperature.

The 2-l beaker containing the main part of the solution is placed in its final position in the water jacket. Its temperature gradually falls to 55°C (for OAM 50°C) and it is held at this temperature for at least 2 days. If the solution is really saturated, crystals will start developing at the bottom of the beaker. If this does not happen the controlled temperature should be reduced rather than start the preparation afresh.

(b) *Preparation of the seeds*

We now turn our attention to the small beaker into which a portion of the solution was put. The temperature of this beaker falls rapidly to room temperature and a crystalline precipitate is produced. This is decanted or filtered off. The remaining liquid, which is now saturated at room temperature, is heated a further 10°C so that it evaporates to a slight extent and then, finally, it is placed in a shallow vessel and left alone. Unless the experiment is spoiled by the formation of a very fine crystalline deposit, perfect seeds develop on the bottom of the vessel. These can be sorted so as to obtain the best.

It takes at least the 2 days, during which the main bulk of the solution is held at the constant initial temperature, to grow seeds of a reasonable size. A few perfect seeds usually develop at the bottom of the beaker, *A*, but these are not available until the growing process has been completed.

The main beaker must not be opened because the slightest sudden drop in temperature produces a fine powder (which we like to describe by the descriptive term 'snow') and this ruins the growth run. These minute seeds eventually fall on the cultivated seeds when these are inserted and twinning occurs. If, for any reason, the 'snow' forms, the temperature should be raised to the initial temperature, vigorous stirring applied and the whole process should be started again.

Five or six of the best seeds are drilled with holes 0·01 inches in diameter and a small nylon thread is attached. Nylon is better than base metals for supporting the seeds because it avoids corrosion. Gold and platinum wires also avoid corrosion but transmit stray vibrations more effectively than nylon. The nylon thread is supported by a float so as to ensure that vibrations are not transmitted from the vessel to the seed. The motor shown in Fig. 10.1 is mounted independently of the whole bath.

When all is ready to insert the seeds—a condition indicated by signs of supersaturation on the bottom of the beaker—the seeds are wetted with distilled water to prevent air bubbles being carried in with them.

When the lid of the beaker is removed, in order to insert the seeds, instantaneous crystallization occurs on the surface of the solution. This layer of crystals sinks to the bottom of the beaker and when the seeds are inserted, the lid replaced and the temperature restored to the previous value, the solution is unsaturated so that the seeds at first slightly dissolve. However, as the temperature is lowered the seeds soon start to grow and within a few hours it is easy to tell if the growing process is proceeding satisfactorily.

It is best to use a number of seeds in any given run because some are always spoiled for one reason or another.

(c) *Factors affecting growth*

(i) *Crystals at the bottom of the beaker*

It is true that the presence of seeds lying on the bottom of the vessel in which growth is occurring reduces the rate of growth of the cultivated seeds. However, it has not proved possible to overcome this difficulty by any form of heating of the bottom of the beaker. The only effect of this was to upset the regular temperature scheme.

(ii) *Creeping of salts*

Whenever experiments were conducted at constant temperature, rather than at constant rate of fall of temperature, great difficulty was caused by the creeping of salts. The salt crept up the wall of the inner vessel, then all over the outer vessel (about 12 in) and then across the table (about 2 ft), then along the bench and finally on to the floor. This difficulty never arose with the method of steady decrease of temperature. It is possible that the condensation on the walls of the inner vessel continually washes any salt back into the solution.

(iii) *Stirring*

The stirring can be always in the same sense—it is not necessary to reverse the direction of rotation. Turbulence must be avoided. In our experiments 'starvation' veils have not been formed. To produce some kinds of good crystals, for example OAM, it is necessary to avoid stirring at all. In one growth of $NaClO_3$ all stirring was avoided to facilitate the taking of a film of the growing process and the crystals were just as well developed as with stirring. When the stirring is gentle, Schlieren streams are easily observable in supersaturated solutions and these show if motion along the growing crystal is satisfactory. The shape of the propeller can be made to correct any defects in the circulation.

The grown crystal must never be taken out of the solution until this has reached room temperature or is not more than 5°C above it. If this condition is not observed thermal stresses are liable to break the crystal.

(iv) *Shaped seeds*

When a crop of good sized crystals has been obtained orientated slices may be cut from them and used as seeds. These slices should have their major faces normal to the direction of most rapid growth, i.e. the major faces should be parallel to the smallest face of the crystal as normally grown. For example a KDP crystal might be grown initially to a size $0{\cdot}5 \times 0{\cdot}5 \times 2$ in. A 45° slice is then cut from it and this will be $0{\cdot}7 \times 0{\cdot}5$ in. When this is used as a seed the cross-section of the resulting crystal will be about $1{\cdot}5 \times 1{\cdot}0$ in. and its length just under 2 in. Such slices, intended as seeds, need not be polished and should be left rough. The cutting may be carried out by a continuous chromel wire carrying carborundum as the abrasive. The carborundum is suspended in glycerine (which does not dissolve the crystal) and fed on to the moving wire before it passes over the crystal. (In large scale production crystals of KDP and ADP up to $2 \times 2 \times 12$ in. may be obtained.)

If the starting temperature is 55°C and the rate of change of temperature is 2°C per day, the whole process will be completed in about 2 weeks.

10.2.3. *Temperature control*

Figures 10.3 and 10.4 show portions of temperature-time records obtained during the process of growth. The measurements cover the temperatures of (a) the solution, (b) the bath, and (c) the room. Drastic changes in room temperature were produced by opening windows in winter time.

Figure 10.3 shows that the temperature fluctuations in the solution are less than those in the bath due to the thermostatic control. The approximate control of temperature of the bath is $\pm 0{\cdot}01$°C. It can also be seen that the duty cycle of the heater is approximately 2 min on and 2 min off. The variations in the temperature of the bath are about ±0·02°C from the set temperature.

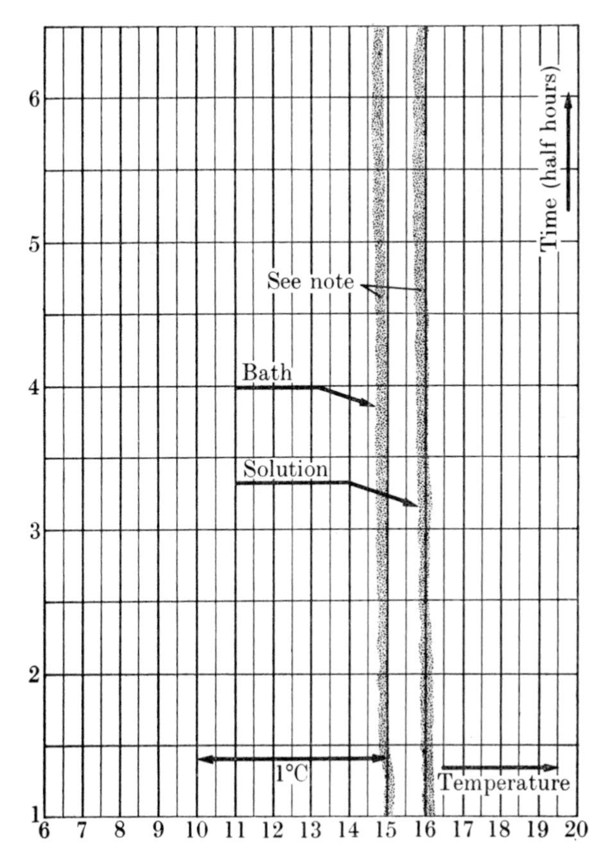

FIG. 10.3. Temperature/time record showing degree of control achieved with the apparatus described.

Figure 10.4 shows that a decrease in the ambient temperature of 5° had no effect on the temperatures of the solution or of the bath. The time taken by the solution to respond to a change in the temperature of the bath was studied. At point *A* on the record the heater was switched off. It took approximately 4 min before there was any change in temperature of the solution. Thus to affect the temperature of the solution any external change must last longer than 4 min. The ordinary variations in the temperature of a thermostatically controlled room will therefore have no effect on the solution if its temperature is not too close to that of the room.

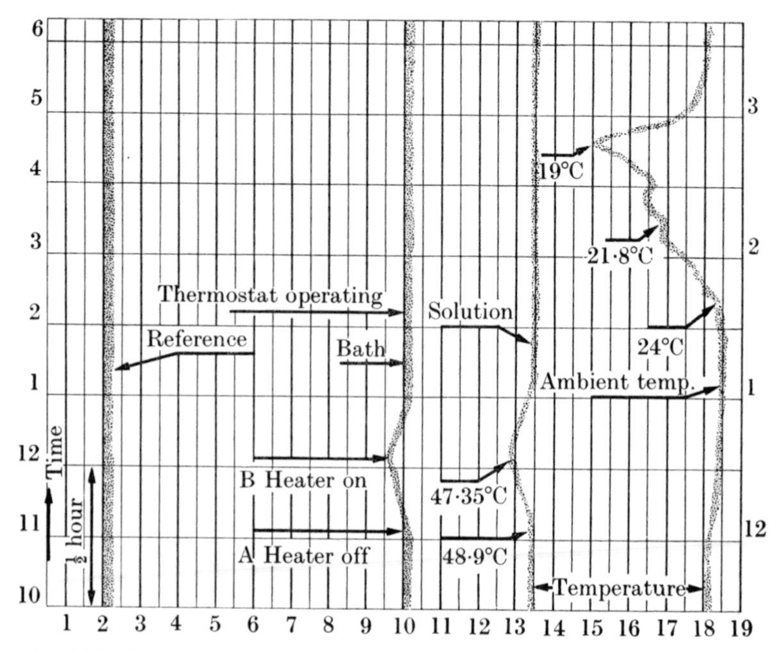

FIG. 10.4. Temperature/time curves showing the effects of various departures from the normal operating conditions.

10.3. Growth from a melt of single crystals of cadmium

10.3.1. *Principle of the method*

Molten material can be grown in the form of single crystals provided that certain conditions are observed. The first condition is that the solidification must commence at only one point. From this point it usually happens that a number of crystals of different crystallographic orientation radiate outwards. The next condition is that the shape of the molten region must be such that only one of the several initial orientations is propagated through the whole of the melt. When the melt is solidifying to a single crystallographic orientation the rate of advance of the boundary between solid and liquid must be controlled so that it never exceeds a speed which is determined by the area of cross-section of the growing crystal. The larger the area of cross-section the smaller must be the rate of advance of this boundary.

10.3.2. *Apparatus for growing single crystal metal wires*

The principles described above are observed in the apparatus for converting polycrystalline cadmium wire into single crystals. The polycrystalline wire is touched on to a heated plate that melts a small piece at the end. The wire is at once withdrawn from the surface so that a more or less conical point is produced on the end of the wire. This end is to be melted and to solidify before the rest and it is from the narrow filament at the end of the conical piece that the single crystal is to begin to grow. The wire with its conical point is put into a glass tube only just large enough to make it possible to insert the wire in it. The function of this glass tube is to prevent the molten material forming a pool of cadmium that breaks up into separate pieces. The tube also has the effect of reducing oxidation. The glass tube, enclosing one piece of cadmium wire, is placed side by side with other similar tubes within a larger glass tube. Before beginning the run the air in the tube is displaced by coal gas or butane and a loose plug of cotton wool is placed in each end of the tube to hinder the re-entry of oxygen from the air. The tube is now returned to the box in which is the travelling furnace. This consists of a heated cylinder some 5 cm long which is just large enough to travel freely along the glass tube without touching it. It is very important that there should be no mechanical disturbance of the growing crystals. Such a disturbance is caused by the furnace rubbing against the glass tube as it passes from one end to the other of the box. Any mechanical vibration of the box is also to be avoided. The furnace is mounted on wheels that run on rails and it is driven by a lead screw operated by a synchronous motor. The temperature of the furnace must be controlled so that it melts a small length of each wire in any given position and it must make this molten region travel along the wire from the drawn out tip to the other end. Fluctuations in the mains potential may or may not affect the result, but the potential applied to the furnace should be kept as constant as conveniently possible. After the furnace has travelled over the wires the larger glass tube can be removed from the furnace. The smaller glass tubes are removed from the larger tube and the wires are shaken out of the smaller tubes into a vessel containing dilute hydrochloric acid. Treatment for a few minutes with the acid removes the oxide skin that may have formed during the melting process. This oxide film greatly reduces the plasticity of the wires.

If the wires appear heavily oxidized the temperature of the furnace should be reduced by lowering the applied potential. If the wires are still polycrystalline after treatment the temperature of the furnace should be raised until it is clear that during the treatment the wire has melted. There are always a certain number of polycrystals produced even when the temperature of the furnace has been correctly adjusted. Frequently these consist of two single crystals that have different crystallographic orientations and form a kind of twin doublet that extends along the whole length of the wire. A single crystal is very soft and if lifted by one end the wire sags under its own weight. Such single crystals must be carefully treated when it is intended to use them for a study of their plastic properties.

11. Phase change and ferroelectricity

11.1. Introduction

WHEN a crystal goes through a phase change, due to thermal expansion, it may be in a polar state at temperatures below the point where the phase change takes place and in a non-polar state above that temperature. In some cases, the polar axis can be reversed by an applied electric field. When this is so the crystal is said to be *ferroelectric* in its polar phase.

Such crystals also show the phenomenon of hysteresis when subjected to an alternating electric field below the transition temperature. The resemblance to ferromagnetic materials is the reason for the name 'ferro-electric'. The law governing the variation of the dielectric constant ϵ with temperature T is

$$\epsilon = \frac{C}{T - T_0}, \tag{11.1}$$

where C is a constant characteristic of each ferroelectric material, and T_0 is the temperature at which the dielectric constant reaches a maximum value (but not an infinite value). The law is named after Curie and Weiss and is known as the Curie–Weiss law. The temperature T_0 is called the Curie point and C is known as the Curie constant. In this experiment C and T_0 will be found for TGS (triglycine sulphate). Below the Curie point dipoles in the crystal are arranged in an ordered manner, the degree of order being affected by the thermal vibrations. In the absence of an applied electric field these dipoles are arranged in a more or less parallel manner in small domains. The polar axis (b-axis of triglycine sulphate) of half of the domains is in the opposite sense to that of the other half, so there is no permanent electric moment in the crystal. Consider an electric field to be applied in the appropriate direction along the b-axis. At first a few domains, which have an orientation of their dipoles opposite to that of the applied field, switch over and become polarized in the same direction as the applied field. As the external field is increased more and more domains switch over until practically all are oriented in the same direction. When this condition is reached the polarization, defined as the electric moment per unit volume, is denoted by P_s and called the saturation polarization. The condition of saturation is represented in Fig. 11.1 by the point A. If, starting from A, the electric field is reduced to zero and then reversed, the corresponding points on the graph of P against E lie on the curve ABC.

If now the field is again reduced and the curve CDA is traversed, we pass through the point E_c at which the polarization is zero and the applied field is E_c. This field is known as the coercive field.

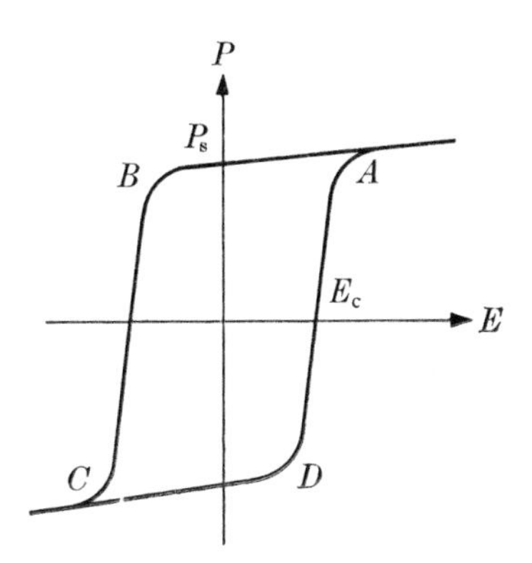

FIG. 11.1. Hysteresis loop obtained on a CRO using a crystal of TGS.

Equation (11.1) may be rewritten

$$T - T_0 = C \cdot \frac{1}{\epsilon}. \tag{11.2}$$

If the Curie–Weiss law holds, then $1/\epsilon$ should bear a linear relation to T and from such a straight line graph the value of C can be found.

11.2. Measurement of saturation polarization and coercive field

The Curie point of TGS is just under 50°C. The crystal is placed in an oven that can be heated to temperatures up to 70°C for the measurement of the saturation polarization, P_s, below the Curie point and also of the dielectric constant, ϵ, below and above the Curie point. Figure 11.2 shows such an oven. Air is drawn from the outside over a heating element and the temperature of the heated air is controlled by a thermostat. Two crystals are placed side by side in the heated space and their temperature is measured with a precision thermometer.

One crystal plate forms one condenser in a 'Sawyer–Tower' circuit, which is given in Fig. 11.3, and this is used to measure P_s. The other crystal is connected to a phase-meter and is used to measure ϵ. In the circuit shown in Fig. 11.3 the 50-Ω, 500-Ω resistors and the associated 1-μF and 0·25-μF condensers are used for filtering the applied alternating potential. The variable 2·5-MΩ resistor and associated capacitors are used to make the phase of the vertical and horizontal deflections of the CRO the same when a normal condenser is put in place of the sample. The oscilloscope trace is arranged to be a straight line under this condition. After this adjustment the normal condenser is replaced by the parallel plate condenser containing the crystal.

The measurement of P_s requires the application of a field that is high enough to reverse the polarization of the crystal. The required voltage

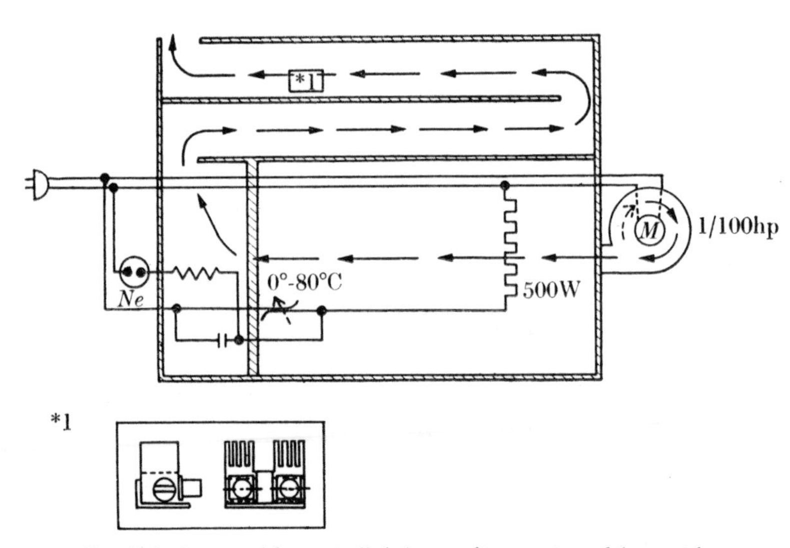

FIG. 11.2. Oven used for controlled change of temperature of the crystals.

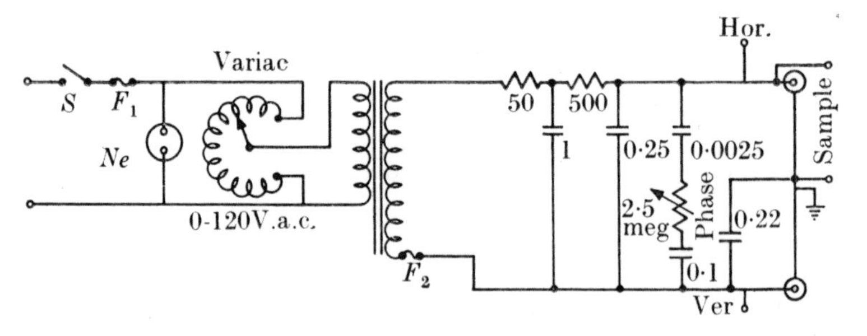

FIG. 11.3. Circuit diagram of the system used in obtaining figures such as Fig. 11.1.

depends on the thickness of the crystal. For instance, in the example used here, which was a plate 1·6 mm thick, in series with a 0·22-μF condenser, 25 V (r.m.s.) was required. Referring to Fig. 11.3 we see that the voltage across the crystal is displayed on the X-axis of the oscilloscope while the voltage across the series capacitor is displayed on the Y-axis.

The value of E_c can be obtained from the X-deflexion and the value of P_s from the Y-deflexion at appropriate points on the hysteresis loop.

The electric displacement, D, the applied electric field, E, the measured dielectric constant, ϵ, the dielectric constant apart from effects due to the electric dipoles, ϵ_0, and the electric moment per unit volume due to the dipoles P, are related by the equations

$$D = \epsilon E, \tag{11.3}$$

$$= \epsilon_0 E + P. \tag{11.4}$$

The reading along the X-axis is proportional to E while that along the Y-axis is proportional to D. The condenser formed by the crystal plate and its electrodes is a parallel plate condenser with a medium of high dielectric constant and edge effects can be neglected. If the charge per unit area of the surface of the crystal is denoted σ we have

$$D = \sigma. \tag{11.5}$$

The charge liberated on the insulated plate of the condenser, Q, is related to σ and A, the area of the plate, by the equation

$$Q = \sigma A. \tag{11.6}$$

Starting from the instant at which the insulated plate is at zero potential we have

$$Q = \int_0^t i\,\mathrm{d}t. \tag{11.7}$$

The 0·22-μF series capacitor must pass the same current as flows through the sample condenser. Therefore the potential across this condenser, V_c, is given by

$$V_c = \frac{1}{C}\int_0^t i\,\mathrm{d}t \tag{11.8}$$

Thus the Y-axis reading of the CRO is proportional to the induction D flowing through the crystal. Because the measurement is dynamic, the moment at which the spot crosses the vertical axis is the moment when $E = 0$. From eqn. (11.4) we can write at this point

$$D = P_0. \tag{11.9}$$

It may be seen from Fig. 11.4 that the slope of the curve is very small at this point and we may put

$$P_0 = P_s. \tag{11.10}$$

If it were not for the term $\epsilon_0 E$ in expression (11.4) the slope of the hysteresis curve would be horizontal along its upper part. The inflexions A, B, C, D in the hysteresis curve show where the domains start to be reversed. When they have all been reversed the slope of the curve should be proportional to the dielectric constant of the material in the absence of reversible dipoles.

The measurements made on the CRO trace are reduced to absolute measure as follows. S_x and S_y are the sensitivities expressed in V/cm along the axes X and Y respectively. When $E=0$ half the vertical distance between the parallel lines AB, CD, Fig. 11.1, is denoted Y. The potential across the condenser C (0·22 μF) denoted V_c, is then

$$V_c = S_y Y. \tag{11.11}$$

From eqns. (11.7) and (11.8) we have

$$Q = CV_c = CS_y Y.$$

From eqns. (11.5) and (11.6) we obtain

$$\sigma = D = CS_y Y/A.$$

Finally, from (11.9) and (11.10) there results

$$P_s = \left(\frac{C.S_y}{A}\right) Y. \tag{11.12}$$

If C is expressed in farads and A in cm², P_s will be in C/cm² or, if we prefer it, in C cm/cm³.

The coercive field is determined in the same way. Let X be half the horizontal distance between the lines BC and AD in Fig. 11.1. Then

$$V_x = XS_x.$$

But

$$E_c = V_x/t,$$

where t is the thickness of the sample.

Thus

$$E_c = S_x X/t. \tag{11.13}$$

It must be remembered that the value of E_c is strongly dependent both on the frequency and on the maximum field applied when the domains are reversed. The actual value of E_c is thus not a significant constant. The shape of the hysteresis curve has more importance.

EXAMPLE

The crystal of TGS was grown as described in Chapter 10, and electrodes were made by painting the opposite faces with high-quality silver paint. The ferroelectric axis of TGS is the b-axis and a cleavage plane is normal to this axis, so it is easy to obtain properly orientated plates. Curves shown in Fig. 11.5 were drawn from data on the photograph of the CRO traces (Fig. 11.4). This data is also given in Table 11.1.

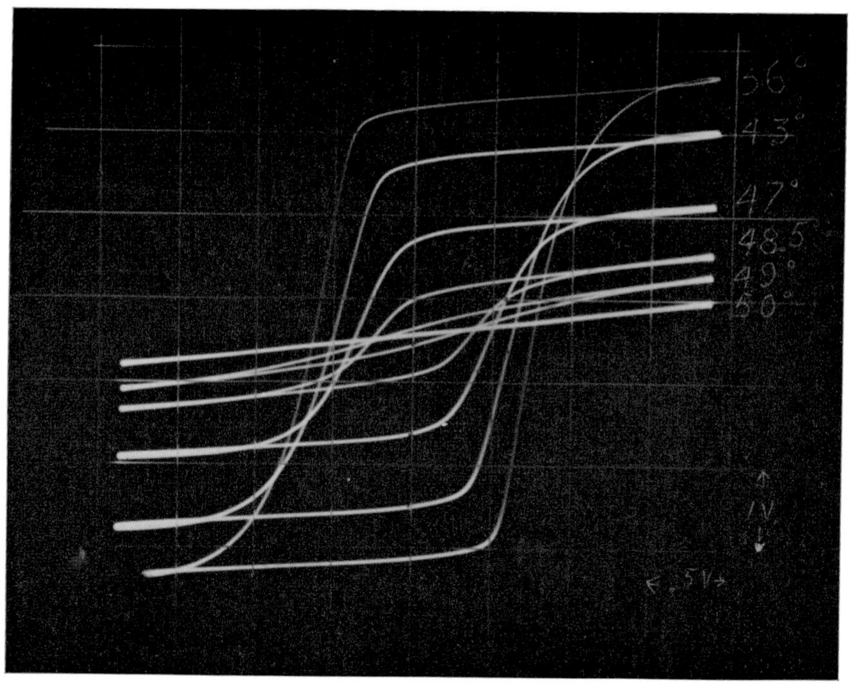

Fig. 11.4. Photograph of hysteresis loops at different temperatures.

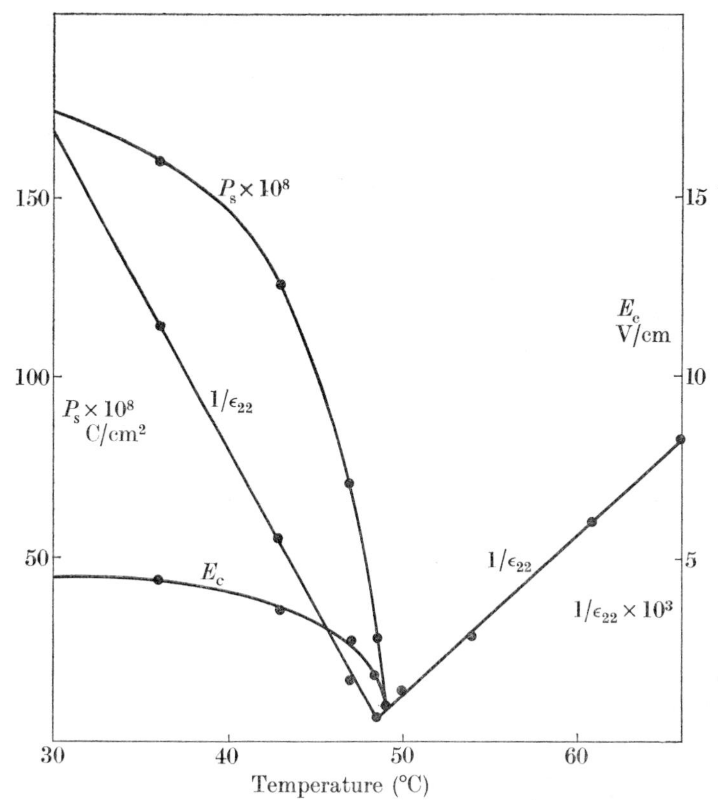

FIG. 11.5. Graphs showing P_s, E_c, $1/\epsilon_{22}$ as a function of temperature for TGS.

TABLE 11.1

Ferroelectric behaviour of TGS along the b-*axis*

Dimensions of plate: 7·3 × 5·0 × 1·6 mm. Series capacitor C: 0·22 μF. CRO calibration: $S_x = 0{\cdot}5$ V/cm, $S_y = 1$ V/cm.

$E_{max} = 11$ V/cm Frequency 60 Hz

Temp. (°C)	Y (cm)	X (cm)	P_s (C/cm² × 10⁸)	E_c (V/cm)	$1/\epsilon_{22}$ × 10³
36	5·3/2	2·8/2	160	4·38	11·55
43	4·2/2	2·3/2	126·5	3·58	5·56
47	2·3/2	1·7/2	69·5	2·65	1·62
48·5	0·9/2	1·2/2	27·2	1·87	0·518
49	0·3/2	—	9·05	—	0·752
50	0	0			1·234
54·1					2·93
61					6·14

As already mentioned, ϵ is determined separately by the method described in Chapter 5 and is taken from Table 5.3.

From Table 11.1 the value of the Curie constant below the Curie point may be found as follows. Taking points from the graph of $T\bigg/\frac{1}{\epsilon_{22}}$ we have

$$C = (48{\cdot}5 - 36)/(11{\cdot}55 - 0{\cdot}518) \times 10^{-3}$$
$$= 1{\cdot}13 \times 10^{3}.$$

11.3. The study of phase-change in barium titanate by observations under the polarizing microscope

11.3.1. *Introduction*

Barium titanate illustrates several features that are characteristic of the changes in physical properties that occur in a number of crystals as their

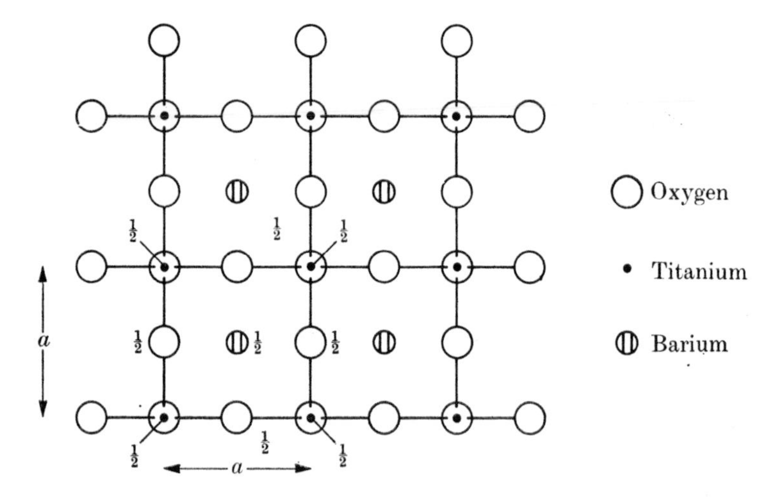

FIG. 11.6. Plan of the cubic crystal structure above 120°C of varium titanate, $BaTiO_3$.

temperature is changed. Above 120°C $BaTiO_3$ is cubic and its structure is represented by Fig. 11.6. The titanium atoms occur at the centres of regular octahedra of oxygen atoms; the barium atoms occur at the centres of the face-centred unit cell of oxygen atoms, and the oxygen atoms form a close-packed arrangement. Below 120°C the crystal becomes tetragonal and its structure is represented in Fig. 11.7. In the direction [001] the former cube edge has become elongated by about ½ per cent and in the other directions perpendicular to [001] it has contracted by about ¼ per cent. All the regularity of the cubic structure has gone. The octahedron surrounding the titanium atoms is slightly distorted, the twelve atoms surrounding the barium atom are not all at the same distance from it, and the oxygen atoms are no longer quite in a regular close-packed assembly. The departures from regularity are not large but they are sufficiently large to introduce new physical properties. The cubic phase is optically isotropic and centrosymmetric so that between crossed Nicols it does not produce the colours characteristic of a birefringent material. The presence of a centre of symmetry means that no piezoelectric effects can be shown by the crystal. When the titanium atoms are all displaced towards the same end of the axis of the surrounding octahedron of oxygen atoms and when the barium atoms have a corresponding displacement in the same direction, the crystal has acquired a polar character. Its class of symmetry is 4 mm so that the

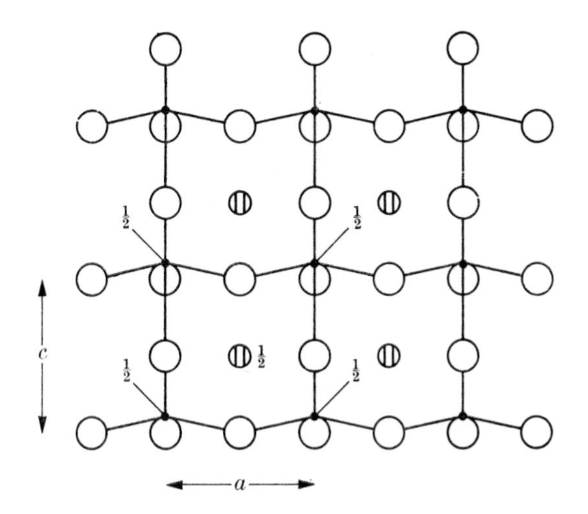

FIG. 11.7. Plan of the tetragonal form between 0 and 120°C of $BaTiO_3$.

tetrad axis is polar. On compression along the tetrad axis the crystal becomes charged positively at one end and negatively at the other. This property of piezoelectricity is combined with the cooperative action of thousands or millions of unit cells all becoming polarized in the same sense and maintaining this spontaneous polarization through the electric field they exert on each other. When an electric field is applied in the opposite sense to that of the spontaneously generated field the whole domain switches to a new orientation associated with this direction of polarization. This reaction is enormously greater than the reaction that would be obtained in the absence of a cooperative effect. There is a hysteresis in the production of the polarization. As the external electric field is increasing the polarization is less than it would be in the absence of a resistance to reorientation of the domains. Similarly, as the external field is decreased the polarization is greater than it would be if there were an immediate reorientation to correspond with the applied field. This hysteresis is due to the influence of the boundaries of the domains that often contain impurities of a chemical or physical kind, including foreign atoms, dislocations, and grain boundaries.

The boundaries of the domains are frequently boundaries between twins. As the cubic phase changes to the tetragonal phase a former cube edge may become the *c*- or the *a*-axis of the tetragonal crystal. Further the $+c$-axis may point in either direction along the given edge. There are thus six possible orientations that the tetragonal phase may take up. Sometimes it forms a single tetragonal crystal but sometimes it forms a complex assembly of domains of varying sizes and orientation.

11.3.2. *Optical observations*

The microscope is of the petrographic type and is provided with polarizer and analyser so that observations can in general be made between crossed polarizer and analyser. Under these conditions a cubic crystal, being isotropic, does not affect the polarized light beam passing through it and consequently no light is seen in the eyepiece. A tetragonal crystal in general alters the state of polarization of the light passing through it, so that the crystal appears bright between crossed polaroids. The intensity and colour of the light depends on the thickness and double refraction of the crystal. Domains that have a tetrad axis normal to the section have zero double refraction; domains with other orientations have a double refraction depending on the inclination of the tetrad axis to the direction of the light. Domains may have different orientations even within the short distance from one point on the lower surface to the point on the upper surface just above it. It is therefore not possible to make a detailed comparison between the domain structure and the optical effects.

11.3.3. *Experimental details*

The apparatus consists of a plate of brass mounted on insulating supports by which it can be carried on the stage of a petrographic microscope, Fig. 11.8. The centre part of the plate is provided with a conical depression in which there is a hole about 2 mm in diameter. The thin plate of $BaTiO_3$ is laid across this hole and viewed by an objective of focal length about 24 mm. The heating is effected by cylindrical elements which may be inserted into holes in the brass plate. A hole is provided so that a mercury-in-glass thermometer may be inserted for measuring the temperature. An insulating layer of asbestos material is fixed to the top of the heated plate to protect the user from burning should he accidentally touch the hot stage.

The heating elements are provided with a controller that can vary the temperature over wide limits. It is best to use full power to begin the heating of the stage and then as a temperature of 120°C is approached to reduce the heating by turning the knob on the controller anticlockwise. By careful adjustment of the controller the optical change can be made to occur slowly and the transition temperature can then be obtained

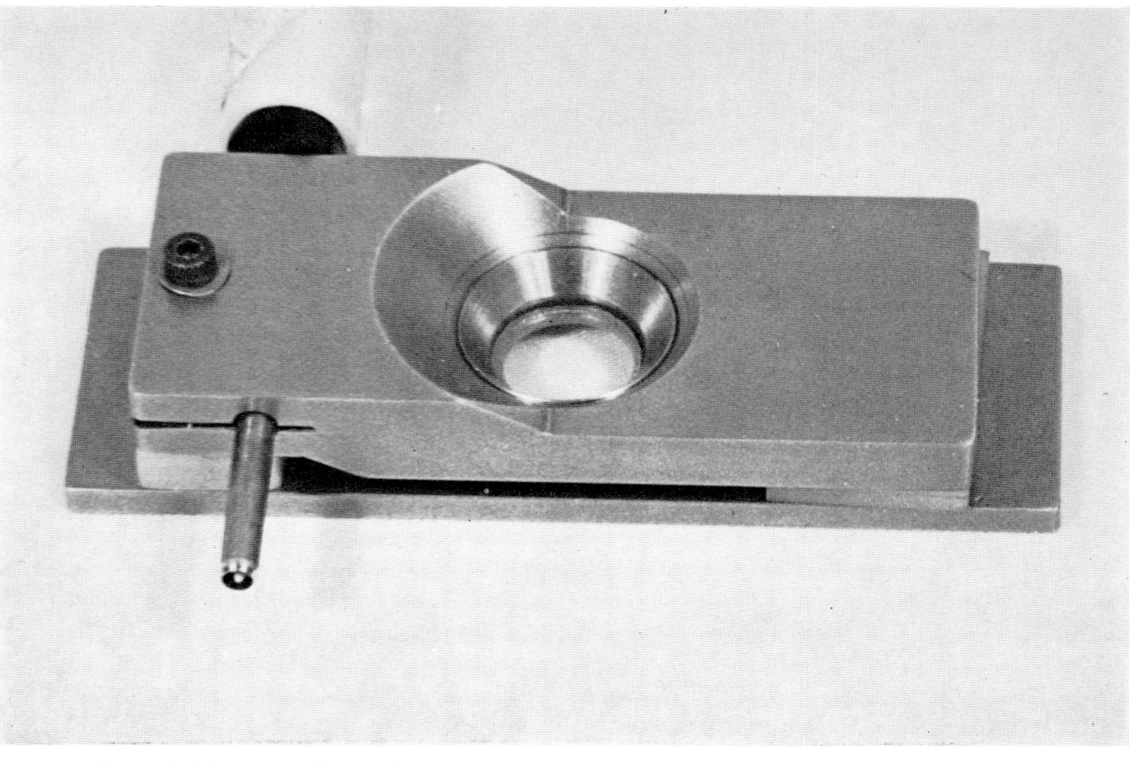

FIG. 11.8. Photograph of a heated microscope stage for the optical study of the phase change in $BaTiO_3$.

reasonably accurately. Measurements should be made both with rising and with falling temperatures.

Note the change from the cubic high temperature form to the doubly refracting lower temperature form. Observe any hysteresis that may be present. Note changes in the form of the figures formed by the changing patterns of domains.

12. Semiconductors

12.1. Introduction

THE Hall experiment, as described in Chapter 8, was the starting point for all the practical applications of semiconductors which are so widespread at the present time. Two features concerning these materials, which are, under different conditions, bad insulators or bad conductors, led to an extension of the theories concerning insulators and conductors. The first characteristic feature is the variation of resistivity with temperature and the second is the existence of both positive and negative coefficients of the Hall effect. The resistivity of semi-conductors, unlike that of typical metallic conductors, decreases with increasing temperature. The Hall effect results require, as was shown in Chapter 8, the existence of *positive* carriers of charge as well as the usual negative electrons.

12.2. Qualitative basis of the theory

It is a fundamental postulate of quantum theory that an electronic energy level cannot be occupied by more than one electron. Experiments involving X-rays have shown the quantized nature of these atomic energy levels. It might be expected, therefore, that when atoms combine together to form a crystal the electrons would be distributed in discrete ranges of energy, which are called bands, and that these would be separated by energy gaps in which no electrons are represented. In the absence of an applied electric field the electrons continue in their various energy levels. When an electric field is applied energy may be given to some electrons and the levels they occupy must be higher on the energy scale than those they previously occupied. If the energy band within which such changes of level might occur is completely filled, i.e. there is one electron for every possible level, then no change in the levels of any of the electrons can occur. Consequently the electric field produces no change of movement of any of the electrons and hence no net current flow. In such a case the material is a perfect insulator.

In conductors the electrons are easily given changes of direction of motion and of energy. This is explained on the present theory by supposing that the last occupied band (corresponding to the last level on the atomic scale) is almost empty. As the temperature is raised, thermal agitation of the atoms increases the probability of collisions between moving electrons and vibrating atoms and the resistivity increases.

The resistance of semiconductors falls as the temperature is raised and these materials should be insulators at the temperature of absolute zero. This implies that at that temperature the last occupied band is completely full and the next to it is completely empty. Electrons can only appear in the empty band if they are elevated from the filled band. This can happen through the action of thermal vibrations that give the energy required by an electron to jump across the gap between the two bands. For this to occur at room temperature the energy gap must be small. When electrons leave the lower band they leave states into which other electrons in the band can be accelerated. In this almost full band there are mutual interactions due to the nuclear fields and accelerated electrons are subject to two fields. One field arises from the neighbouring atoms and the other is the applied field.

An analogy with motor traffic in an urban street may be helpful. There are two extremes: in one there are only a few cars, which occasionally interfere with one another's movements and, in the other case, the street is almost jammed full of cars but there are some empty spaces between them. When there are only a few cars in the street their movement to the right constitutes a flow in that direction. When the street is nearly full a flow of spaces to the left corresponds to the flow of an equal number of cars to the right. This analogy attempts to illustrate the fact that a flow of electrons in one direction gives the same transfer of charge as a flow of an equal number of holes in the opposite direction. When electrons have jumped from the almost filled valency band the holes provide a means for the acceleration of electrons but it is simpler to regard the flow as being due to an opposite flow of positive charges.

This type of theory explains why the resistance decreases as the temperature rises. The greater thermal agitation provides for more electrons to jump out of the valency band and so provides for conduction in both bands.

The theory next takes account of the possibilities that impurity atoms may be added and that these may be N-type (donors of electrons) or P-type (acceptors of electrons). When an impurity atom has one electron more than its neighbouring atoms in the crystal structure, this one electron is very loosely held and may easily jump out of the valency band. This increases the concentration of free electrons in the material and such a material is called a *majority-carrier* of the negative type. Alternatively, the impurity atoms may be electron acceptors having one electron fewer than their neighbours in the crystal structure. Valency electrons in this case make a jump to occupy the missing band. In this case we have a majority-carrier of the p-type.

The fact that the Hall effect sometimes corresponds to the carriers being predominantly positively charged, supports the theory outlined above.

Before going further, recourse must be had to some simple arguments based on conditions of thermal equilibrium. The following quantities are represented by symbols: p, p_i stand for the concentration of holes in the impure and intrinsic (pure) materials respectively; n, n_i stand for the corresponding concentrations of electrons; g, g_i represent the rates of generation of electron-hole pairs in the impure and intrinsic materials; and r, r_i represent the corresponding rates of recombination.

The rate of production of free electrons depends only on temperature since it is the thermal agitation that provides the energy to jump over the potential barrier. Thus we have

$$g = g_i. \tag{12.1}$$

In the intrinsic material when a free electron is produced a hole is also formed so that

$$p_i = n_i. \tag{12.2}$$

Under conditions of thermal equilibrium the rate of generation of holes or free electrons must always equal the corresponding rates of recombination. If this were not so charge of one sign would accumulate continuously. Thus,

$$g = r,$$

$$g_i = r_i. \tag{12.3}$$

From eqns. (12.1) and (12.3) we obtain

$$r = r_i. \tag{12.4}$$

The rate of recombination must also depend on the probability of collisions between holes and electrons. This probability must be proportional to the product of the concentrations so that

$$r = Knp,$$

$$r_i = Kn_i p_i. \tag{12.5}$$

Combining (12.4) and (12.5) we have

$$np = n_i p_i = \text{constant}. \tag{12.6}$$

From (12.2) this constant is n_i^2.

Thus by adding *N*-type impurities while n is increased relative to n_i, p is *decreased* with respect to p_i. In fact it does not require much impurity to make the concentration of minority carriers almost negligible in comparison with that of the majority carriers.

12.3. The junction of *p*- and *n*-type materials

It follows from the considerations of the previous section that a junction between *p*- and *n*-type materials has certain special properties. Figure 12.1 indicates the state of affairs. (Note that the concentrations are expressed on a logarithmic scale in order to bring all the features to be considered within the one diagram.) At the junction there exists a high concentration of electrons on one side and a very low concentration on the other. The same is true for the concentrations of holes. Diffusion occurs across the junction just as in gases. The electrons from the *N*-side diffuse into the *P*-side. The majority concentration *Nn* on the *N*-side is not depleted relatively as much as the minority concentration on the *P*-side is increased. Only a thin layer at the junction is depleted and this is called the transition region. The process of diffusion eventually stops because as the electrons diffuse from

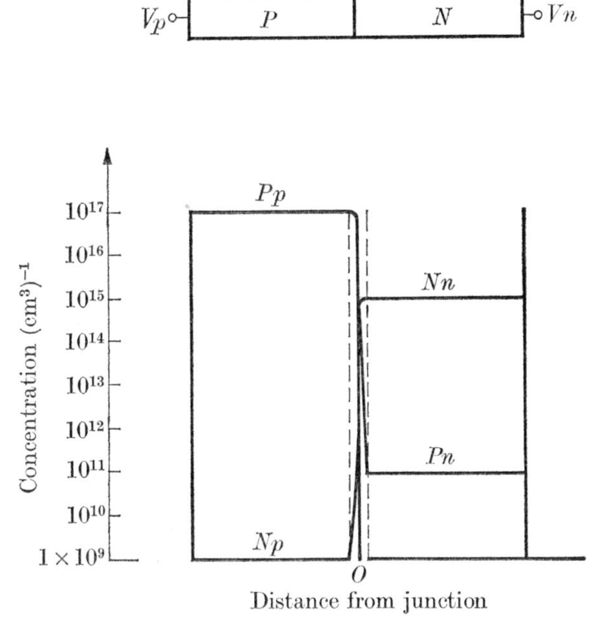

FIG. 12.1. Diagram showing the concentrations of positive and negative carriers in semiconductors of *P*- and *N*-types as a function of distance from the junction. The left-hand letter of each pair of letters refers to the sign of the carrier. The second letter refers to the type of semiconductor at the given point of the diagram.

N to *P* they leave behind positively charged fixed ions and also increase the negative charge on the *P* side. The junction becomes polarized with the *N*-side positive with respect to the *P*-side. The corresponding migration of holes from the *P*-side produces negatively charged fixed ions (neutral atoms that have accepted or trapped an electron) on the *P*-side and adds to the positive charge on the *N*-side. Thus, eventually, all migration stops. Figure 12.1 represents the state in the neighbourhood of the junction when migration has stopped. It should be remembered that in the absence of an externally applied potential no current through the junction is possible.

When a so-called forward voltage is applied, making the *P*-side more positive with respect to the *N*-side, the potential barrier is being reduced and this encourages further diffusion. If the external d.c. potential is steadily applied it prevents the full potential barrier being built up and a steady diffusion current flows. Minority concentrations at the junction are increased with respect to equilibrium values and the excess over the equilibrium values increases exponentially as the applied voltage increases (see Fig. 12.2).

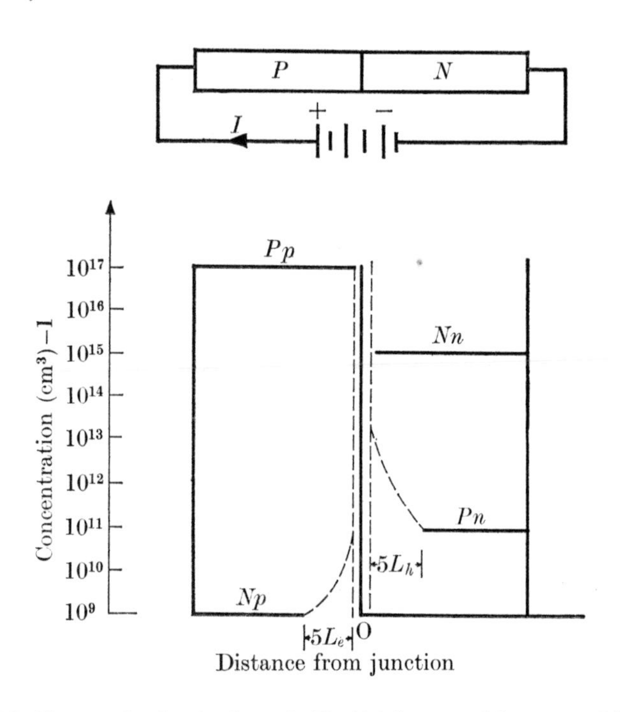

FIG. 12.2. Diagram showing the change in Fig. 12.1 due to applying a potential across the junction. The distances marked $5L_e$, $5L_h$ represent the ranges over which the concentration of electrons and holes respectively are disturbed by the applied potential difference.

On the other hand, when the junction is reverse-biased the applied potential increases the effect of the natural potential barrier. Apart from collecting carriers that are due to thermal agitation on either side of the junction, no current can flow. Minority concentrations at the junction are decreased with respect to their equilibrium values.

When a forward bias is applied negatively charged electrons flowing in the negative direction of *X* form a positive current and positive holes flowing in the positive direction of *X* also produce a positive current. Both diffusions are additive and give a positive forward current. The resistance of a forward biased junction is very low and if the potential barrier is completely overcome the externally applied voltage becomes equal to the voltage drop along the semiconductor bar which depends only on its resistivity.

12.4. The mechanism of gain in a transistor

Figure 12.3 represents an arrangement of two diodes arranged so that one is forward-biased and the other back-biased. As was explained in section 12.3, this will merely result in current i_1 being very high and current i_2 being very small. This arrangement of junctions can be produced in one bar, making the first region *P*-type, the next *N*-type, and next *P* again. This arrangement becomes useful when the second junction is made so close to the first one that it is situated within the $5L_h$ region shown in Fig. 12.2.

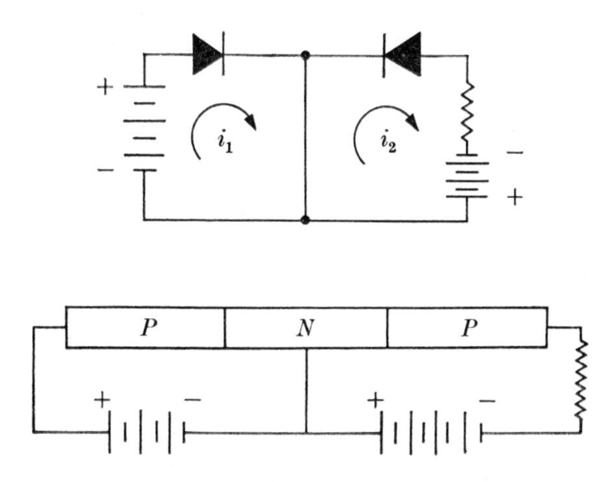

FIG. 12.3. Diagram of a circuit containing oppositely directed diodes.

At the reverse-biased junction the excess carriers emitted at the forward junction are not recombined with their opposites but are collected and flow through into the *P*-type section. The closer the second junction is to the first the more nearly are all the transmitted carriers collected. The first *P*-type material is called the emitter, the *N*-type material is called the base, and the second *P*-type material is called the collector. The concentrations of holes and of electrons when the proper potentials are applied to such a transistor are shown in Fig. 12.4.

The arrangement described above provides a *voltage gain*. This follows from the forward bias on the emitter-base junction which makes it have a low resistance and the reverse bias on the base-collector junction which makes it have a high resistance. If a load resistor is placed between the source and the collector a voltage gain is possible provided the load resistor is higher in value than the resistance of the emitter-base junction.

If the emitter-base junction is forward-biased and a small signal is fed in at the base a current gain is possible. In this case the signal source has only to supply the recombination current. Typical values of the emitter-collector current gains are 0·98 to 0·99. The recombination current to be produced at the base is thus from 0·02 to 0·01. The ratio of the emitter-collector to base-collector current is thus from 0·98/0·02 = 49 to 0·99/0·01 = 99.

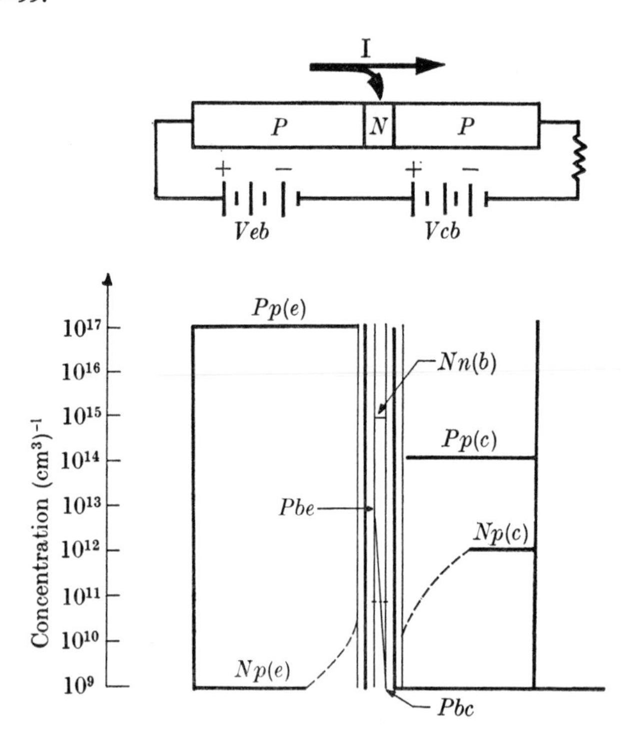

FIG. 12.4. Diagram showing the effects obtained when the middle layer is sufficiently thin. The letters *e*, *b*, *c* stand respectively for 'emitter', 'base', and 'collector'.

12.5. Experimental procedure

A circuit suitable for measuring the transfer characteristics of transistors is shown in Fig. 12.5. An alternating voltage is applied between collector and emitter so that at any given time one complete characteristic curve can be seen on the CRO. A diode is placed between the secondary of the transformer supplying the power in order to cut out the irrelevant half of each cycle. A battery supplies the base current which is varied by means of the 50-kΩ potentiometer. A switch S_1 opens or closes both the mains supply (117 V a.c.) and the internal 9-V battery. A small load resistor is used to measure the collector current. Lastly, a main switch reverses both the 9-V battery and the series diode D_1 so that either *PNP* or *NPN* transistors may be used. The diagram refers to measurement of *NPN* transistors.

The voltage between the emitter and the collector is applied to the horizontal axis of the CRO. The length of the horizontal trace is adjusted to be within the screen and then the scale factor is determined by applying

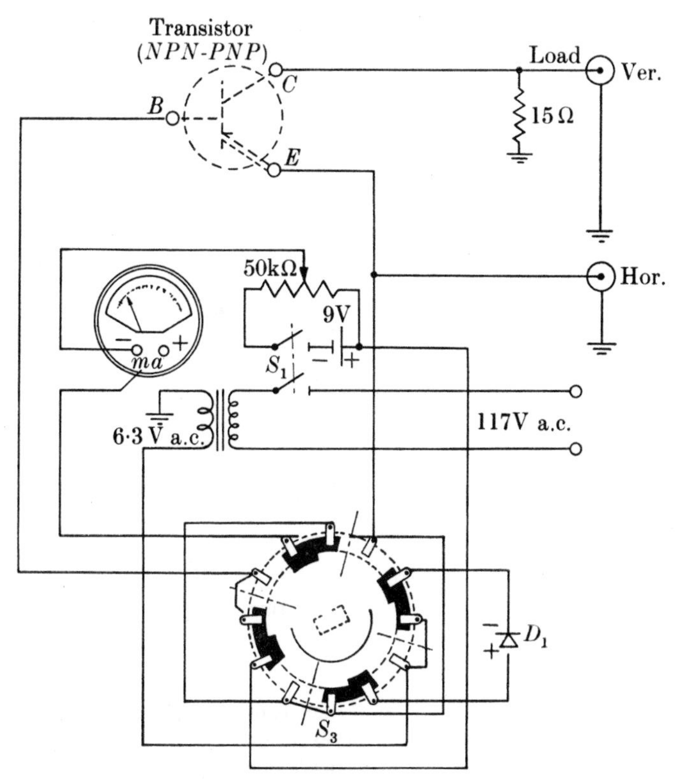

FIG. 12.5. Circuit diagram of the arrangement used for obtaining the current/voltage curves for the collector of a transistor using various base currents.

the same voltage to the vertical deflexion. This is usually calibrated directly in V/cm.

The voltage across the load resistor is applied to the vertical deflexion. The vertical gain should be adjusted so that the trace is on the screen for the maximum value of the base current, I_b, the current flowing into the base. The voltage across the load is readily obtained from the vertical deflexion and, knowing the resistance of the load, the collector current I_c can be found.

I_b is varied manually in equal increments and the corresponding curve is traced on the oscilloscope. If a camera is available, a photograph should be taken of each curve of the family (see Fig. 12.6) in order to obtain the complete family on the same photograph. While photographing each curve of the family, the graticule should not be illuminated. Once the complete family is recorded the vertical axis is disconnected in order to record the horizontal axis. Finally both axes are disconnected, the graticule is illuminated and recorded. In this way the picture is calibrated and positioned.

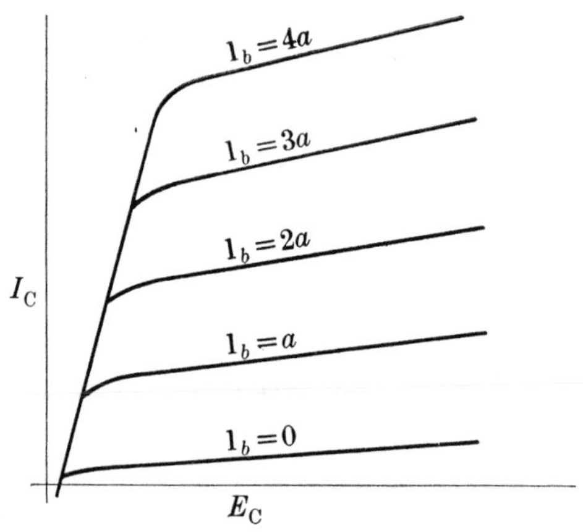

FIG. 12.6. Diagram of current/voltage curves obtained from apparatus shown in Fig. 12.5.

EXAMPLES

Figure 12.7 shows the actual photograph obtained with a polaroid camera. The transistor used was a *PNP* germanium power transistor (RCA 2N 2869).

Calibration of oscilloscope

The full length of the horizontal trace is 6·8(3) divisions and this corresponds to 4 V. This deflexion therefore corresponds to 0·586 V/div.

The vertical deflexion was set to 0·2 V/div. The load used was 15 Ω so that in measuring the collector current the calibration is $0{\cdot}2/15 = 13{\cdot}3$ mA/div.

Base current

There are six traces in Fig. 12.7 corresponding to base currents 0, 0·2, 0·4, 0·6, 0·8, and 1·0 mA respectively.

Characteristics obtained

(1) *Short circuit current gain*, β, in the nearly horizontal parts of the curves:

$$\beta = \frac{\Delta I_c}{\Delta I_b}.$$

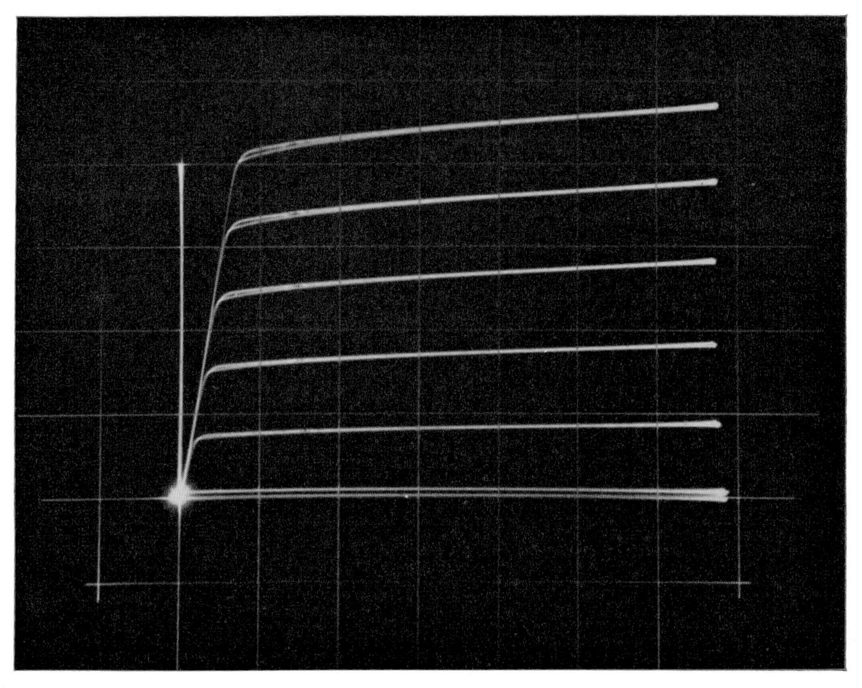

FIG. 12.7. Photograph of the CRO traces obtained with the circuit of Fig. 12.5.

Along the fifth vertical line to the right of the origin the separation between successive nearly horizontal traces is 1 div., i.e. $\Delta I_c = 13{\cdot}3$ mA and $\Delta I_b = 0{\cdot}2$ mA. Thus

$$\beta = \frac{13{\cdot}3}{0{\cdot}2} = 66{\cdot}5.$$

(2) *Open circuit transfer impedance*, r_{21}, is obtained along the third horizontal line of the grid above the origin. For this line

$$I_c = 3 \times 13{\cdot}3 = 39{\cdot}9 \text{ mA},$$

$$r_{21} = \frac{\Delta E_c}{\Delta I_b}.$$

Between the sixth and the first vertical lines of the grid the value of $\Delta E_c = 2{\cdot}9$ 3V. The corresponding values of I_b are 0·64 and 0·72 mA. Thus

$$r_{21} = -2{\cdot}93/0{\cdot}08 = -37.$$

(3) *Output resistance*, r_d, is obtained along a constant I_b curve

$$r_d = \frac{\Delta E_c}{\Delta I_c}.$$

Between the first and sixth vertical lines of the grid to the right of the origin $\Delta E_c = 2{\cdot}93$ V. For $I_b = 0{\cdot}6$ mA, the value of ΔI_c is one-third of the spacing between two grid lines, i.e.

$$\Delta I_c = 0{\cdot}3 \times 13{\cdot}3 \times 10^{-3} \text{ A}.$$

Thus

$$r_d = \frac{2{\cdot}93}{0{\cdot}3 \times 13{\cdot}3 \times 10^{-3}} = 734\ \Omega.$$

(4) *Operating conditions*

If a 75-Ω load were used at an E_c value of 4 V the load line, which gives corresponding changes of voltage across and current through the transistor would have a slope of $1/75 = 0{\cdot}0133$. Starting from a point $I_c = 0$, E_c 4 V, this will cut the I_c axis at a point $4 \times 0{\cdot}0133 = 0{\cdot}0532$ above the origin. Each grid unit corresponds to 0·0133 and hence the point of intersection would be four grid lines above the origin. Thus the range over which operation without distortion of the signal is possible is from I_b 0 to 0·82 mA giving a peak-swing of 45·2 mA. This corresponds to an r.m.s. value of 16 mA. The proper bias would give $I_b = 0{\cdot}41$ mA so that the quiescent I_c would be 22·6 mA and the collector d.c. bias 2·34 V. ($4 \times 0{\cdot}586$).

The values 45·2 and 22·6 mA are obtained from the intersection of the load line with the I_b curves for 0·82 and 0·41 mA respectively.

(5) *Power dissipation*

The load of 75 Ω dissipates:

$$\text{d.c. power} = 75 \times (22{\cdot}6 \times 10^{-3})^2 = 0{\cdot}0383 \text{ W},$$

$$\text{a.c. power} = 75 \times (16 \times 10^{-3})^2 = 0{\cdot}0192 \text{ W}.$$

The transistor dissipates:

$$\text{d.c. power} = 2{\cdot}34 \times 22{\cdot}6 \times 10^{-3} = 0{\cdot}053 \text{ W},$$

$$\text{a.c. power} = 734 \times (16 \times 10^{-3})^2 = 0{\cdot}188 \text{ W}.$$

$$\overline{\underline{0{\cdot}241}}$$

The d.c. supply must provide the total of 0·3 W.
The only useful part is 0·02 W.
The total dissipated in the transistor is 0·241 W.

Note. In this example the transistor was not used at its full capacity. A 734-Ω load resistor would be required for maximum power transfer and for this a much higher bias voltage would be necessary. When working under full load considerable heat would be developed and a heat sink with fins to assist the cooling would be necessary.

13. The measurement of resonance characteristics of ferrimagnetic materials†

13.1. Introduction

FERROMAGNETIC metals and alloys are not generally useful at frequencies above a few megahertz because of their usually high conductivity. Ferrimagnetic materials, on the other hand, have the double advantage of a very low conductivity and negligible dielectric losses. Many ultra-high frequency devices are based on the properties of ferrimagnetics. In their high-frequency applications, ferrimagnetic materials, namely, ferrites and garnets, are often used at resonance. We shall therefore, in this experiment, study them at or near to resonance. From these experiments fundamental properties needed in the design of many devices can be obtained.

For the conduct of these experiments strong steady magnetic fields and ultra-high frequency electromagnetic radiation is required. Steady fields of a few kilogauss can be obtained with usual electro-magnets and the frequency of the ultra-high frequency radiation must then lie in the range 8–12 GHz. To generate, control, and measure such radiation klystrons, attenuators, and special meters are required. With this type of apparatus it is possible to measure the resonant frequency of the magnetic ions contained in the ferrite, the band width of the resonance and hence the magnetic losses. It is also possible to determine an important parameter known as the effective Landé g-factor and the saturation magnetization.

13.2. Theoretical background

The magnetic moment associated with an electron is attributed to its spin and its orbital motion around a nucleus. When a magnetic field is applied to it a couple acts in the same way as gravitational force acts on a spinning gyroscope supported at one end of its axis of rotation. Just as a gyroscope precesses about a vertical axis under these conditions so the magnetic moment of an electron precesses about the direction of the applied magnetic field. Electromagnetic radiation can be generated which has the same frequency as this frequency of precession, known as the Larmor frequency. When such radiation falls on a magnetic ion there is a strong interaction between the radiation and the magnetic moment of the ion. This is a similar phenomenon to that which we encountered in the piezoelectric resonator. The energy absorbed by the precessing moments in

† The authors are indebted to Prof. P. Lavallée, Ecole Polytechnique, Montréal, for his contribution to this chapter. Prof. Lavallée supplied all the experimental data.

resonance with the incident electromagnetic radiation is large and can be measured by the reaction on the source of supply of that radiation. The relation between the resonance frequency ω_0, and the steady magnetic field strength, H_0, is, for a spherical sample,

$$\omega_0 = \gamma_e H_0, \tag{13.1}$$

where γ_e is a constant which in MKS units can be shown to be

$$\gamma_e = -\mu_0 g \cdot \frac{e}{2m} \tag{13.2}$$

where μ_0 is the permeability of free space ($4\pi . 10^{-7}$), g is the effective Landé splitting factor, and e and m are the charge and mass of the electron. If there were no energy loss in this resonance the breadth of the resonance peak would be vanishingly small. In fact it varies from one material to another.

A graph is plotted showing the power of the electromagnetic radiation absorbed by the specimen as a function of the strength of the steady magnetic field. This graph contains a resonance peak and two points on it are found which are half-way between the level on either side of the peak (background level) and the peak itself. The distance between them corresponds to a change ΔH in the strength of the magnetic field and this quantity increases as the energy loss at resonance increases. A coefficient α is defined by the equation

$$\alpha = \frac{\Delta H}{2H_0} \tag{13.3}$$

where ΔH is the so-called half-height line-breadth and H_0 is the applied field at the centre of the peak.

13.3. Experimental arrangement

Figure 13.1 shows the general arrangement of the component parts of the necessary apparatus. From left to right these parts are as follows.

(1) A reflex klystron with its d.c. power supply. This tube oscillates at ultra-high frequencies in the range 8–12 GHz because its principle of operation (velocity modulation and bunching) avoids the limitations inherent in transit time tubes like the triode,

(2) The ferrite isolator which is generally used to prevent frequency 'pulling', that is the effect a varying load impedance would otherwise have on the output frequency of the klystron. The isolator is usually of the resonance type. A thin magnetised slab of ferrite, suitably positioned in the wave guide reacts differently to waves travelling in opposite directions on account of the different sense of rotation of their respective r.f. magnetic fields. Thus an outgoing wave in passing through the isolator is hardly attenuated. A reflected wave travelling back towards the klystron is attenuated by resonance absorption. Some 20 dB is a typical isolation,

(3) The frequency meter is a tunable cylindrical cavity coupled to the wave guide through a hole in the broad face of the latter. At resonance, the cavity absorbs and dissipates in its walls a small fraction of the power in the main line. This sudden reduction of power is detected and as the position of the piston is calibrated in terms of the resonance frequency, the latter may be read off directly. Once the frequency is determined the cavity is detuned.

(4) the attenuator which is a variable resistive load in series in the line —it can be adjusted to control the amount of energy reaching the sample,

(5) the directional coupler is a secondary line coupled to the main line with coupling holes so spaced that no power enters it from the incident wave but the reflected wave is able to enter it. The power transferred from the reflected wave to the derived branch of the directional coupler is

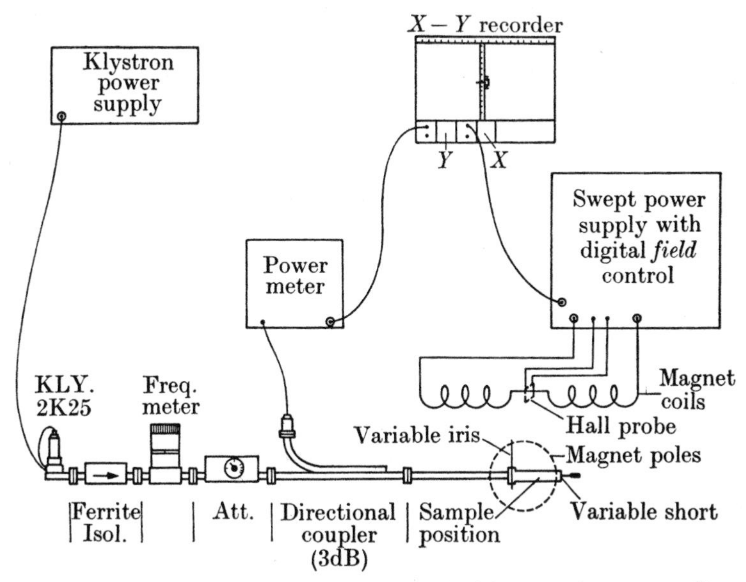

FIG. 13.1. Diagram of the apparatus used in the study of the magnetic resonance effects in ferrites.

detected and measured by a power meter, the reading of which can be transferred to the y-axis of an X–Y plotter.

(6) the variable iris diaphragm couples the sample cavity to the wave guide. The size of the opening depends in general on the size of, and losses in, the sample. Typically the diameter of the hole is about $\frac{1}{8}$ inch while the thickness of the diaphragm is $\frac{1}{32}$ inch.

(7) The sample cavity is rectangular. The sample should be spherical so that the simple equation (13.1) can be used to relate the frequency to the field at resonance; otherwise demagnetization factors would have to be

used explicitly. This sample is mounted in a small block of foamed plastic. This has a relative dielectric constant close to unity and, unlike solid supports, does not distort the r.f. field round the specimen. It is very important to position the sample in the cavity so that it is where the r.f. magnetic field is strong and perpendicular to the slowly varying magnetic field produced by the electromagnet. The diameter of the sphere is usually about 0.5 mm.

(8) an electromagnet capable of supplying a magnetic field that can be swept through a certain range that includes the value corresponding to resonance within the sample. The strength of this magnetic field is read by a Hall probe and transferred to the x-axis of an X–Y plotter. The magnetic field is swept through the value corresponding to maximum absorption by the sample and the resonance curves such as those shown in Figs. 13.2, 13.3 are plotted automatically.

EXAMPLES

(1) *Yttrium-iron garnet* $Y_3Fe_2(FeO_4)_3$

A sphere of this material was ground accurately spherical for this experiment. The record given by the X–Y plotter is shown in Fig. 13.2.

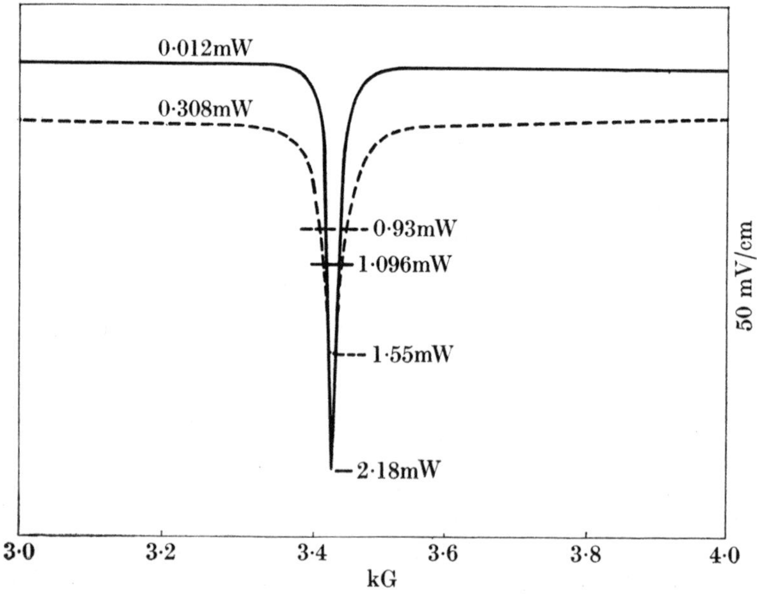

FIG. 13.2. Curves obtained by plotting power absorbed in ferrite versus the slowly varying magnetic field for a sample of yttrium iron garnet $Y_3Fe_2(FeO_4)_3$.

The ordinates give the power absorbed and the abscissae the steady magnetic field strength. Two curves (full and dotted) show the effect of changing the size of the iris diaphragm. It will be seen that the peak is narrower with the larger opening. The vertical scale of the graph corresponds to the sensitivity (50 mV/cm) of the plotter. The frequency of the radiation produced by the klystron was 9·75 GHz which corresponds to

$$\omega_0 = 2\pi . 9{\cdot}75 \times 10^9 \text{ Hz}.$$

From the graph it may be seen that the centre of the absorption trough corresponds to a magnetic field of 3·44 kG. We may therefore write

$$H_0 = 0{\cdot}344/\mu_0 = \frac{0{\cdot}344}{4\pi . 10^{-7}} = 274\ 000 \text{ A/m}.$$

From eqn (13.1)

$$\gamma_e = \omega_0/H_0 = 2{\cdot}24 \times 10^5.$$

From eqn (13.2) we obtain

$$g = 2\frac{m}{e}\frac{\gamma_e}{\mu_0} = 2{\cdot}03.$$

The line width is given by

$$\Delta H = 0{\cdot}016 \text{ kG}$$
$$= 0{\cdot}0016 \text{ Wb/m}^2.$$

From eqn (13.3) we obtain

$$\alpha = \frac{\Delta H}{2H_0} = \frac{0{\cdot}016}{2 \times 3{\cdot}44} = 0{\cdot}00233.$$

(2) *Magnesium polycrystalline ferrite* (ferramic R-1)

The experimental results are shown in Fig. 13.3, from which the following results can be obtained.

$$H_0 = \frac{0{\cdot}316}{4\pi \times 10^{-7}} = 251\ 000 \text{ A/m}.$$

$$\omega_0 = 2\pi \times 9{\cdot}75 \times 10^9 \text{ Hz}.$$

$$\gamma_e = \frac{\omega_0}{H_0} = 2{\cdot}44 \times 10^5.$$

$$g = 2{\cdot}2$$

$$\Delta H = 0{\cdot}132 \text{ Wb/m}^2.$$

$$\alpha = \frac{\Delta H}{2H_0} = \frac{0{\cdot}132}{2 \times 0{\cdot}316} = 0{\cdot}209.$$

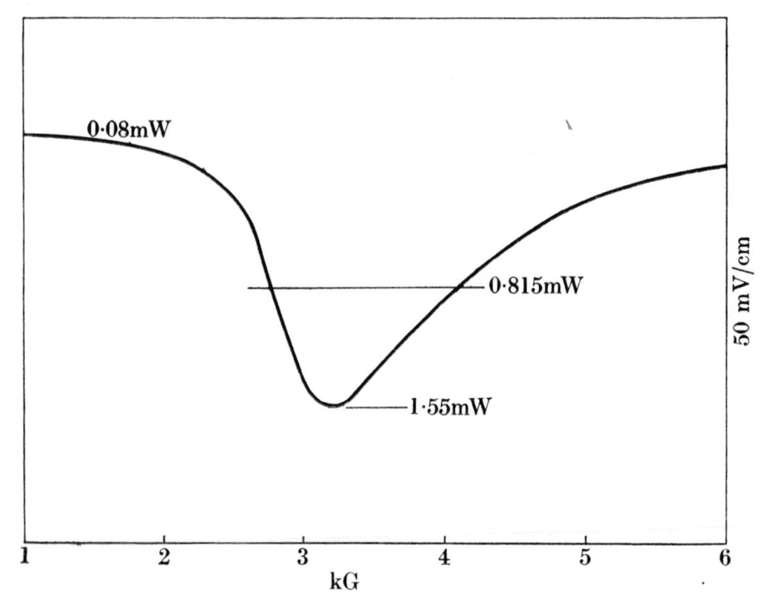

FIG. 13.3. Corresponding curve to those of Fig. 13.2 obtained with a polycrystalline magnesium ferrite (ferramic R-1).

Bibliography

BLOSS, F. D. *An introduction to the methods of optical crystallography.* Holt, Rinehart, Winston; New York, London (1966).

BUCKLEY, H. E. *Crystal growth.* Wiley, New York; Chapman and Hall, London (1951).

CADY, W. G. *Piezoelectricity.* McGraw-Hill, New York (1946).

DANA, E. S. and HURLBUT, C. S. *Manual of mineralogy.* Wiley, New York (1941).

EWALD, P. P., PÖSCHL, TH., and PRANDTL, L. *The physics of solids and fluids.* Blackie, London (1932).

JONA, F. P. and SHIRANE, G. *Ferroelectric crystals.* Pergamon Press, London (1962).

DE JONG, W. F. Compendium der Kristallkunde. N. V. A. Ooshoek's *Uitgevers-Maatschappij.* Utrecht (1951).

KATZ, H. J. *Solid state magnetic and dielectric devices.* Wiley, New York (1959).

KAYE, G. W. C. and LABY, T. H. *Physical and chemical constants.* Longmans, Green, London (1949).

KLEBER, W. *Einführung in die Kristallographic.* VEB Verlag Technik, Berlin (1956).

LIEBISCH, TH. *Physikalische Krystallographie.* Verlag von Veit, Leipzig (1891).

MASON, W. P. *Piezoelectric crystals and their applications to ultrasonics.* D. Van Nostrand, New York (1950).

MEGAW, H. D. *Ferroelectricity in crystals.* Methuen, London (1957).

MIDDLEBROOK, R. D. *An introduction to junction transistor theory.* Wiley, New York (1957).

NYE, J. F. *Physical properties of crystals.* Clarendon Press, Oxford (1957).

PHILLIPS, A. H. *Mineralogy.* Macmillan, New York (1912).

PHILLIPS, F. C. *An introduction to crystallography.* Longmans, Green, London (1956).

PUTLEY, E. H. *The Hall effect and related phenomena.* Butterworths, London (1960).

SCHUBNIKOV, A. V., FLINT, E. E., and BOKY, G. B. *Elementary crystallography.* Academy of Science USSR, Moscow and Leningrad (1940).

SEITZ, F. *The modern theory of solids.* McGraw-Hill, New York and London (1940).

SMIT, J. and WIJN, H. P. J. *Ferrites.* Philip's Technical Library, Cleaver-Hume Press, London (1959).

TUTTON, A. E. H. *Crystallography and practical crystal measurement,* Vol. II. Macmillan, London (1922).

VOIGT, W. *Lehrbuch der kristallphysik.* Teubner, Leipzig and Berlin (1928).

WINCHELL, A. N. *Elements of optical mineralogy.* Wiley, New York (1948).

WOOSTER, W. A. *Crystal physics.* Cambridge University Press (1949).

List of Original Papers

BAUMHAUER, H. *Z. Krystallogr. Miner.* **3**, 588 (1879).
DEBYE, P. and SCHERRER, P. *K. Ges. Wiss. Göttingen*, Dec. 1915; *Phys. Z.* 1 July 1916, p. 277.
DE SENARMONT, H. *C.r. hebd. Séanc. Acad. Sci., Paris*, **25**, 459, 707 (1847); *Annln. Phys.* **73**, 191; **75**, 50, 482 (1848).
FORBES, J. D. *Phil. Trans. R. Soc. Edinb.* **23**, 133 (1862).
FRESNEL, A. *Oeuvres Compl.* **1**, 731; *Annls. Chim. Phys.* (2) **28**, 147 (1822).
GAUGUIN, F. M. *C.r. hebd. Séanc. Acad. Sci., Paris*, **42**, 1246 (1856).
GIEBE, E. and SCHEIBE, A. *Z. Phys.* **33**, 760 (1925).
HOSHINO, S., MITSUI, T., JONA, S. and PEPINSKY, R. *Phys. Rev.* **107**, 1255 (1957).
JONA, F. and SHIRANE, G. *Ferroelectric crystals*, p. 30. Pergamon Press, London (1962).
KOHLRAUSCH, F. *Annln. Phys.* N.F. **4**, 1 (1878); **16**, 609 (1882).
KRISHNAN, K. S., GUHA, B. C. and BANERJEE, S. *Phil. Trans. R. Soc.* **A231**, 235 (1933).
LAVAL, J. L'état solide (Rapports et Discussions), p. 273. Congress Solway Bruxelles, Stoops (1951).
MARTIN, A. J. P. *Mineralog. Mag.* **22**, 519 (1931).
MAURICE, M. E. *Proc. Camb. phil. Soc. math. phys. Sci.* **26**, 491 (1930).
RÖNTGEN, W. C. *Z. Krystallogr. Miner.* **3**, 17 (1879).

Subject Index